# The Possibility Dogs

BOOKS BY SUSANNAH CHARLESON

*Scent of the Missing: Love and Partnership
with a Search and Rescue Dog*

*The Possibility Dogs: What a Handful of
"Unadoptables" Taught Me about Service,
Hope, and Healing*

# The
# Possibility Dogs

WHAT A HANDFUL OF "UNADOPTABLES"
TAUGHT ME ABOUT SERVICE,
HOPE, AND HEALING

## Susannah Charleson

*Mariner Books*
*Houghton Mifflin Harcourt*
BOSTON    NEW YORK

*The Possibility Dogs* is a memoir grounded in events and interactions that took place across a decade. Out of respect for individual privacy, some names and identifying circumstances have been changed.

First Mariner Books edition 2014
Copyright © 2013 by Susannah Charleson

www.hmhco.com

*Library of Congress Cataloging-in-Publication Data is available.*
ISBN 978-0-547-73493-4   ISBN 978-0-544-22802-3 (pbk.)

Book design by Melissa Lotfy

Printed in the United States of America
DOC 10 9 8 7 6 5 4 3 2 1

PHOTO CREDITS

*Insert page 1:* Susannah Charleson (top, middle left), Devon Thomas Treadwell (middle right, bottom); *insert page 2:* Susannah Charleson (top), Devon Thomas Treadwell (bottom); *insert page 3:* Susannah Charleson (top, bottom right), Devon Thomas Treadwell (bottom left); *insert page 4:* Susannah Charleson (top left, top right), Devon Thomas Treadwell (bottom); *insert page 5:* Alex Washoe (top left, top right), Susannah Charleson (bottom); *insert page 6:* Susannah Charleson (top left), Shauna Galligan (top right, bottom); *insert page 7:* Susannah Charleson (top left), Carlos Salcedo (top right), Raymond J. Benoit (bottom); *insert page 8:* Susannah Charleson (top, bottom right), Devon Thomas Treadwell (middle), Ellen Sanchez (bottom left)

*For assistance dog Logan,*
*beautiful golden gone far too soon,*
*and for Suzan and Jon Morris,*
*who cherish her still*

# The Possibility Dogs

# 1

IF HE CHOOSES TO REMEMBER, what he remembers first is the taste of grit in eggs. Cold scrambled eggs on a blue plate with a half-moon chip along the edge. He had the eggs sometime that morning, he's pretty sure, probably with toast, probably with coffee, though it's only the eggs and the plate he can still get hold of — he makes parentheses with his hands — the battered plate and the rubbery eggs gritted with sand, leftover sand from a goddamn sandstorm that had blown hard the day before, rusting the sky and coating everything and everyone with a layer of red grime. It was the kind of sandstorm that would make you think the world was ending, but it wasn't. Next day the sun came up like any other, only you could write *wash me* on every surface of the house and eat dirty eggs for breakfast.

If he chooses to remember, he repeats, that's what he gets.

Everything else is flash and bang. Bob knows mostly what other firefighters have told him. There was a fire that afternoon. A vacant two-story building almost a hundred years old, for the record sitting empty, but broken windows, bedrolls, and graffiti inside said vagrants had been living there awhile. A convenience store clerk across the street smelled something before there was anything to see, and then there was a haze of smoke and a lick of flame, and that old building went right up. The clerk made the call. The fire department was there fast, trucking in water to this nonsupplied area, but the fire blazed upward two beats ahead of them, smoke rolling from the windows, distorted walls bulging and bowing from the pressure, the gush of water pink where it met yesterday's sand.

The whole building was involved, and containing the fire was the issue, there being a row of other derelict buildings on the block — some houses nearby, too, with families in them — equally old and primed like kindling. The firefighters were on it, and on it, and on it some more, putting out the fire only to have it spring up again in some far corner. A company from a nearby town also responded. After a couple of hours, it was over, the building three-quarters gone, soggy and smoldering. Some of the men had tried to search for trapped victims early in the blaze, but it had forced them back. Now that the fire was spent, it was time to go in.

The ugliest job of all. They were thigh-deep, pushing through the shit of it, he's been told, finding no living and no dead, when a triangle of roof collapsed. Bob doesn't remember that roof, but the roof brought down a wall that he does remember: a roar of falling brick, and he was running toward a slash of sunlight, several of his company steps ahead. This memory repeats like a loop tape. It is a race he thinks they'll win every time. He went down in the joy of escape and woke buried — how long later? — to the exhausted wail of a friend dying three feet away.

He was in the hospital a long, long time.

A man born to motion, he has a stillness about him now. Bob is burly, obviously powerful — tall and broad without being fat — eclipsing the wooden chair that holds him. But when he speaks, there is vulnerability, a sense of bewilderment, as though something here a moment ago, and precious, is gone. A big guy with a ravaged face and a wavering voice, he's tough to read. You could be scared by him or moved. Bob knows this. I'm *difficult,* he says, turning a little more toward me from his position at a table flanked by windows. He reaches down to touch his dog. She looks up into his eyes even before the touch connects.

They are an unexpected pair sitting quietly together in this tapas bar in Baltimore. The salt-and-pepper speckled dog at Bob's left, just in the place where he can drop his hand and cup the dome of her head, is petite. She is thirty pounds, maybe, with delicate, deerlike white feet. The rest of her is spots all over. A circle of rust-colored fur around one eye looks like a monocle. She could be a Border collie crossed with an Aussie shepherd whose mother had had a fling with a husky fathered by a beagle. A right mix. Her red vest reads *In Service* on one side and *Please Do Not Distract* on the other.

When the host first sat me at the table next to them, the man looked up, nodded, and looked back down at his menu. The dog gave me a single glance, sweetly alert, a double huff of nostrils that said she'd tagged my scent, and then she turned away. This dog was all business. When her partner cleared his throat, she looked up into his face immediately, holding her gaze there until something in his behavior satisfied her, and then she relaxed again. He extended his index finger to her cheek and gave it an absent scritch. She opened her mouth in the half smile, half pant of a happy dog. They obviously had things worked out.

I partner a search-and-rescue dog of my own, so I was naturally intrigued by the pair. Since I've been on the other side of the admiring glances and the outright stares in public, I turned away, giving them their privacy. But the seated man had seen the logo of the search team I worked with, the silhouette of a dog, embroidered on my shirt, and he started the conversation.

"Canine search-and-rescue?" he asked. I nodded. "Retired firefighter," he said about himself, pointing his fork toward his chest. "Worked with the search dogs a few times. Good work."

"Beautiful partner you have there."

"Haska." The dog glanced up at the sound of her name, poised forward on her keel and elbows in question, as if she was ready to get up if he was. He scratched her cheek, and she settled. "Bob." He pointed the fork to his chest again.

"Susannah."

We could see through the wide windows that it was still a nasty day here in Baltimore — strong winds scudding across a low-hung sky; intermittent sleet following a week of snow. Now and then, sun sliced through the gray, then disappeared again. Bob and I were the only patrons in the restaurant, both of us here for conferences. His hotel was rumored to have a norovirus running amok; my hotel had only one restaurant to serve several hundred attendees who didn't want to leave the hotel and brave the ice.

Both Bob and I had cabin fever, it turned out, so today we headed here from separate directions, teetering across the slick sidewalks to the Inner Harbor, to find lunch. We will end up sitting here for hours.

We have much in common, we discover as we talk across two tables. He was born in Texas; I live there now. I'm a flight instructor; he took a dozen lessons a long time ago. He was a firefighter who sometimes worked with search dogs; I am a search dog handler who sometimes works with firefighters. We explore that shared ground, about calamity and scent and where the goodwill of the dogs instructs the lesser parts of us, and looking down into Haska's face, Bob says that his dog has never seen a fire that he knows of, but she saves him plenty every day. That might sound sentimental from anyone else, but from him, it's frank. *This is my dog. This is what she does.*

Bob clears his throat often. He speaks in fits and starts — a burst of information and then a full stop for half a minute or more. He tells me about the fire, about losing fellow firefighters in the accident. I was lucky for so long, he says. Bound to happen sometime — getting hurt, losing brothers. Traumatic brain injury and subsequent psychological conditions put him in the hospital for almost a year. He stayed with a sister for another year after that.

Once he was able to live independently again, Bob decided to get a dog. He says he wanted something to come home to, a dog that needed him as much as he needed her. Bob has balance problems now and then, so he didn't want a small dog he might stumble over and crush, and at the city pound he nearly rejected petite Haska on that count. But he liked how she came right up to him when he approached her kennel and put her paws up on the chain link so softly that they didn't make a sound. He fell for her pretty fast, he says, but because she'd been taken in as a stray who might have an owner searching for her, Haska wasn't adoptable when they first met. So he went back every day to visit, and on every visit Haska wiggle-butted her way to him across the cold cement. Shelter staff took pity on their mutual devotion, and Bob was able to put in a first-bid application for her even before she was available. No one came to claim her, and Haska became adoptable on a Wednesday at 9:00 A.M. Bob got her at 9:01.

And then she was his, and he was hers, and everyone, *everyone* questioned it, Bob says, because he's recovering from an injury that should go away and posttraumatic stress disorder that maybe never will. He has headaches often and nightmares sometimes and tinnitus, which causes some everyday noises to be unbearable. He once walked into a quilting shop with his mother and the sound of a sewing machine nearly took the top of his head off. The worst thing about all of it, he tells me, is that he can't go back to firefighting, ever, and yet he seems to live in a state of emergency.

It was his therapist who suggested a service dog trained to intervene during PTSD events, something Bob had never heard of. They were discussing his ability to financially support a second dog when they realized he already had a trainable dog at home who could take on the job. Haska is now more than just a sensitive pet that is devoted to her owner. Through careful training, she has learned the changes of voice and posture and the fidgeting that begin her partner's episodes. Always near Bob, at these times she is trained to redirect him. A nudge of his hand or a lick of his fingertips is sometimes enough, but in nightmare season, sometimes she has to paw him awake. He gets flashbacks, sometimes, and the public shakes, as he calls them, often, and then she may have to repeat her correction behaviors a dozen times or more: *Here I am, here you are, here we are together.* Haska can guide him from buildings. If he gets disoriented on neighborhood walks, she can lead him home.

She's very patient, but she's also insistent. Bob says it's Haska who grounds him in a better reality. Not every bang is a falling wall. A squeal of brakes is not a scream. He looks to his dog to tell the difference. Bob clears his throat again when he says this, a tense, habitual grind. Haska looks up at him thoughtfully, and he puts his hand to her head. In that moment, I see that the throat-clearing must be something he does both in an episode and out of it, and her assessment and his immediate response is the quick check between them.

Haska is a psychiatric service dog, the first I have ever met. Her carefully developed partnership with Bob is a hard-won achievement on both sides, one that allows him to lead a fuller, independent life. Bob is proud of his dog and proud of himself, but he says they can't really rest on their laurels. His disability, while thoroughly documented, is invisible. Any public outing has potential conflict. He takes a sip of iced tea and guesstimates that every third restaurant they enter tries to throw them out.

According to the Code of Federal Regulations, a service animal is any animal that is individually trained to do work or perform tasks for the benefit of an individual with a disability. The Americans with Disabilities Act defines a disability as a mental or physical impairment that substantially limits one or more of a person's major life activities. Regulations specify that the tasks performed by a service animal must be directly related to his handler's disability. Disabilities may be physical, sensory, psychiatric, intellectual, or combinations of any of these, and dogs can be trained to serve them all.

It is mighty work. There are thousands of service dogs performing thousands of tasks, partnering children and adults in public and private spaces worldwide. Most recognizable, perhaps, are guide dogs serving the blind; they began to be used in the late 1920s. Guide dogs stepped into the public consciousness in the years following World War II, when many veterans returned from battle with vision impairment, and the need for assistance was high. These partnerships were recognizable by both the harness and vest on the dog and the obvious disability of the handler.

Guide dogs have proved their worth over decades. Other kinds of canine service have also demonstrated their importance: there are hearing-assistance dogs, dogs that aid mobility and balance, dogs that pick up or carry needed objects for their partners, dogs that respond to seizures, alert diabetic crises, or other catastrophic medical conditions, and, much more recently, dogs like Haska, who are trained to intervene in their handlers' psychiatric disorders by performing specific tasks to prevent, lessen, or redirect their behaviors. Talented service animals empower human partners in all kinds of ways. But despite the service dogs' long history, a great deal of media attention, the protections offered by the Americans with Disabilities Act, and formal statements made by the Department of Justice, problems still arise. Misconceptions about them are common, everyday errands can create conflict, and in public places, partners with service dogs are often wrongly shown the door.

Why is this? The United States provides some of the greatest support for service dog partnerships in the world, but perhaps problems still occur because service animals' full access to public establishments was not a right until the Americans with Disabilities Act made it so, in the late twentieth century. Up to then, the public lives of assistance-dog-and-handler teams were limited. Now, more than twenty years later, some proprietors still don't know about ADA protections. Some are not aware of the difference between a pet and a service animal and perceive any dog on the premises as a health-code violation. Some believe the dog's presence is a safety risk to other customers and that they, the proprietors, are liable for any injury or accident that might occur. Some don't understand the nature of canine service or believe in it. Some suspect they are being duped by ordinary pet owners. Others openly admit they simply don't like dogs. Finally, especially in the case of medical-response canines and those that serve handlers with invisible disabilities, it's not merely the necessity of the dog that's questioned but also the existence of the disability itself. And for these partnerships, some of the greatest problems arise.

In some ways, Bob says, he is fortunate. He is a big man and imposing, and as he was a firefighter, he is confident in tense situations and used to taking charge. Sometimes he doesn't get questioned, and he attributes this to his size and the firefighter insignia he has on a couple of jackets. He's been given a discreet high sign a few times, the wink-wink, as though proprietors were willing to sneak in the two of them under the halo of emergency services, as if Haska were a disaster dog and Bob were getting a quick hamburger before they went back on the job. This was especially true in the early years following 9/11. In some ways, this made things easier for Bob, and he rolled with it, and in other ways, he says, it was frustrating. Bob had the uncomfortable feeling that sliding in as a firefighter wasn't going to help the next guy who came along with a service dog.

And so he has tried to become an educator, he says, which means being far more open about himself and his condition than he could have ever imagined three years ago. Bob has taken off the firefighter jacket. He has learned to recognize the double take at his dog, and the hesitation, and when he's questioned, he responds by saying, "This is my service dog, Haska. She is trained to prompt me out of PTSD episodes that I can't predict." Bob recognizes that in telling others his dog's task, he necessarily discloses the disability that Haska serves. He is not required by law to state his disability to anyone. But how can he explain what she does without revealing something of himself? He's not a wordsmith. He shrugs and says he thinks this is a place he had to get to. It's all right. Strangely enough, the more he owns his condition, the more empowered he seems to become.

Bob and Haska leave the restaurant before I do. I see them out the door, hear the *ting-ting* of Haska's tags as she pads with him down the stairs. Moments later, they move cautiously across the reddish pavement on the water side of the Inner Harbor, bright figures on a gray day in early February. They are completely in sync. I like my last view of them, the big man and the petite dog picking their careful way across the ice.

# 2

I UNDERSTOOD BOB BETTER than I let on to him. While I've never experienced disaster in the ways that he had as a firefighter, I know some of the symptoms he described all too well. I've had the nightmares and the breathless wakefulness that follows them, the sense of universal urgency, and the insomnia that leads to punchy, disconnected days. Construction bangs and squealing brakes don't bother me, but I do know what it is to be caught in a set of behaviors that seem almost impossible to control.

In 2004, many months after taking part in an ugly search that affected me more than I immediately knew, I was diagnosed with critical incident stress by one mental-health practitioner and with PTSD by another. The terms are essentially related, though they aren't synonymous. Posttraumatic stress disorder has received a great deal of press and become better known in recent years. But what is critical incident stress? A critical incident is defined as a life experience that seriously upsets the balance of an individual, creating changes in the way that person functions emotionally, cognitively, or behaviorally. Critical incident stress, like PTSD, can occur in victims of crime or warfare, in abuse survivors, and in those who serve in crisis response or in battle. For those working in the heart of disaster, among the most common contributors to critical incident stress are serious injury or jeopardy, line-of-duty death or suicide of a colleague, the death of a child, a failed rescue attempt, mass casualties, and fatalities of people known to the responders. For police, firefighters, EMTs, paramedics, and med-flight personnel, potential trauma is out there almost every day.

I suppose that, with me, it was difficult to tell where the CIS left off and the PTSD began. The tags put on my condition didn't matter as much to me as its effects. What I knew was that I sometimes felt too much — depressed, anxious, guilty — and then weeks of relative numb would follow. I sought counseling because search-and-rescue training teaches us to recognize such symptoms, but they showed up so long after the initial event, I couldn't make much sense of them. I wasn't sure how long what I was feeling would last or what I should do, or if there was anything to do at all. Because I hadn't missed any work, I falsely made light of it, telling one counselor it was uncomfortable but survivable, calling it my "portable gloom." I didn't confess the one image that repeated itself in dreams and sometimes in my waking thoughts, repeated so profoundly that one morning a surge of grief dropped me to my knees in front of the kitchen sink, and I was surprised to find myself there.

You think you are prepared, of course, when you head out on a search and get advance warning that it's going to be ugly. This one would be bad — human catastrophe at its most vicious, across rough and remote terrain. I steeled myself for it and went in to do the job like everyone else, prepared to look squarely at the human death we were likely to find. But despite all that steeling, I wasn't prepared for everything.

I wasn't prepared for the dogs. As we bounded out beside the well-fed, lucky search K9s, we often passed many local dogs that were victims of neglect. Anxious and angry, left out on chains at the edge of some properties, these were hungry dogs with infected pressure sores, the contours of their ribs and hipbones visible beneath patchy coats. There seemed to be many of them. A resident of the area told me that dogs were often used as crime deterrents and that the common belief was that a hungry dog was more savage. Savage dogs were highly prized, so some didn't get fed a lot. We saw such dogs often, felt them watching us while we worked. Not every dog there was in bad shape, of course, but we saw some terrible cases.

One dog touched many of us who came into contact with her. At the edge of a yard bordering a parking lot used by search personnel, there was a brindle-and-white pit bull on an eight-foot chain. She was horribly thin, with sunken eyes, and she appeared to have pupped recently, her teats descended and breasts collapsed as though spent. The first morning we parked there to head in for the day's briefing, we couldn't

help but see the bones of her as she stood beneath her owner's No Trespassing sign. She strained forward on the end of her chain, growling
sometimes but never barking, her nose working the scent of us and,
surely, the scent of dog food that many of the canine handlers stored in
their cars.

Her emaciated state was impossible to ignore. Between search de-
ployments over the course of several days, some of us snuck fast-food
chicken to her from a place across the street. Several handlers started
tossing her a cup of kibble a few times a day. The poor girl seemed to
inhale it, her tail whip-whipping as she gulped down the food. Then
one morning, a sign was posted:

NO FEEDING THE GODDAM DOG

Which seemed to ensure she'd get fed more.

The canine handlers did a little investigating. They inspected her
part of the yard to make sure the feeding wasn't causing digestion
problems; no, I heard a couple of them decide — this dog was being
starved for the hell of it.

Someone should be told, they said.

Someone should Do Something.

What something?

Feed her, certainly.

Take her. On the second day after the sign appeared, that muttered
suggestion slid across the hoods of cars and pooled at our feet, unre-
solved, a conflict of conscience, law, first duty to the search, and the
possible ramifications for the law enforcement agencies overseeing us
— as well as the real possibility that even a fed dog could bite.

Much was said. The local authorities were told. Yes, yes, they knew
this dog, but their response was unpromising, and we could feel them
turn away. *You have jobs to do,* we heard in their dismissal. That dog
wasn't the reason we were here. First duty to the search. They were
right in that, of course, but the pitbull remained a figure of conscience
for many of us. We saw her every day, and we left her every day looking
hopefully after us.

On one chilly afternoon, I was paired to search with a sheriff's deputy who worked with a beautiful Belgian Malinois. We were to drive

to a remote area miles away and meet a bus of community volunteers. They would grid-walk the large search area with us and a couple of other canine handlers and cadaver dogs. My partner and I were first to head out. We had either bad directions that led us to the wrong place or good directions we didn't follow well, and we made at least two false stops. The Malinois was the only one happy about the long, bumpy drive in light sleet. She huffed over my shoulder, straining forward to see the view.

Finally, the deputy and I took a last turn into the woods, stopped the truck, got out, and waited for the bus to arrive. We had a hand-drawn map as well as a scribbled description of the area we were to search; it was uninhabited and ambiguously owned, and although we were told to keep our eyes open, search officials didn't think we'd encounter any problems there. We were stiff with cold and eager to get started. We paced a little, huffing steam into our cupped hands to warm our faces. Enough waiting already. The officer left his dog in the truck, and we crunched our way across the turf frosted with ice pellets, thinking we would get a visual on the search boundaries and see just how big this area was.

But the terrain didn't seem to match up to the description we'd received. Our notes could have described a hundred places out here, really. Two-lane country road: check. Worn, unpaved turn to the south: check. We crossed a creek bed, saw woods and unimproved land on the left, barbed-wire fence on the right. That fence petered out to nothing, collapsed with disregard. Landmarks noted on the map were sort of there and sort of not. We followed the dirt road about two hundred yards, walking down a long slope into the quiet wood.

For all its isolation, this place had seen some traffic. Vehicles had worn the road smooth — a number of vehicles over time, each following the route of countless others. I suppose that should have alerted one of us. Why all the traffic out here in the middle of nowhere? We found the churn of tire tracks in the lowest spots, dug deep, slicing the old mud in wide half circles, as though after a hard rain, several trucks had gotten stuck or come here for a little off-road fun.

There was no sign of anyone here now. And where were the volunteers? Where was the bus? Where were *we*, exactly, with regard to this hand-drawn map? The world seemed far too quiet even for a rural area. If there were local birds, we had spooked them into silence. We crunched on for a while, saying nothing.

The smell of death was upon us before we saw it, a funk rising up out of nowhere and falling so heavily that all the other odors of the woods gave way to it. There was something odd about it, and while I knew immediately this wasn't going to be who we were searching for, I also knew it was going to be bad.

"Oh shit," said the deputy ahead of me. He had stumbled into cinder blocks and old chain link. "Jesus," he said when he came upon the dogs.

A lot of dogs. Thirty or forty of them, maybe. Prisoners all, and dead. Some small dogs were huddled in low, patched-together kennels of wood and chicken wire. Other, bigger dogs lay in slightly larger runs of chain link over pea gravel, beneath an unsteady framework of wood and corrugated tin. The little dogs were housed two and three together. The larger ones seemed to have been kept by themselves.

Labs, pit bulls, hounds, goldens, terriers, spaniels — and a lot of fuzzy, indeterminate mixed breeds. Some of the dogs had died miserably pressed against the wire; others lay curled in fetal, hopeless positions. A few had been worked by carrion eaters. Anything could have killed the dogs in those kennels — starvation or sickness; exposure, perhaps. A small dirt pit about twenty yards away held the charred remains of three dogs. They lay apart, stretched out as though they were running. They appeared to have been torched.

I don't know how long I actually stood there — long enough to take in those desperate faces — but it couldn't have been much more than a minute. I turned so abruptly that I ran into a tree, witless. The young deputy scrambled away a few feet behind me.

We got out of there as fast as we could, struggling back the way we'd come. The deputy stopped once and bent over with his hands on his knees, and I thought for a moment he would vomit. I looked away and rubbed my hands to stop their shaking, feeling grief, fury, helplessness — guilt, even. Why guilt? But guilt was there, so strong I can feel it still.

What had we seen? Was this a fighting facility? A puppy mill's crude kennel? A cult's sacrifice operation or a hoarder's sick idea of rescue? Whatever it was, these dogs had died apart from mercy of any kind.

When we moved on, I was quiet. The deputy was not. Getting angry seemed to help. I envied him his threats and ugly promises, hacking away at the memory, like it was something to be felled. We slogged out of the mud and ran down the long road we never should have taken in the first place.

We worked on, in that sector and others; we worked on because that was our job, making terrible finds that were part of the search. And then the search stood down. So small the order — a crackle of static and an "All-call, return to base," and the news spread across the area like a stain. Search personnel came in from the nearer spaces and the far ones and gathered for the debrief, mute with finality. Afterward, I saw the young deputy across the room. His eyes were rimmed red with fatigue, but he bobbed his head at me and mouthed the word *Dogs*, then pointed to himself and a local officer.

Packing up, heading out, I saw the neglected pit bull one last time at the edge of the parking lot. She still rushed forward on her chain, but she had begun wagging her tail when cars rolled up to park. She was wiggly now, hopeful in the way some dogs can impossibly be hopeful after hardship. *Don't wag at us, baby*, I thought. *Don't wag*. I could imagine the punishment that wag might earn. Someone had crossed out the sign's NO and ING and then turned the sign around to face the owner's house. FEED THE GODDAM DOG, it read. It was a rant, a vent, and it wouldn't change a thing.

As much as I wanted to save her, I didn't. I got in the car and I left her. In the rearview mirror, her eager face faded from sight. That image carried as I drove away from that town, through another and another and another. Unable to escape my conscience and my cowardice, I finally pulled the car over and cried. I had never cried after a search and haven't since, but I needed to shake this one off. I needed to shake off the pit bull, too, and breathe free of the memory of those tortured dogs, but I could still smell their suffering and see my partner trembling with rage. "What the fuck? What the fuck?" the young man had cried, frustrated with our universal impotence at saving anything at all.

So when I couldn't escape my own sadness and spoke to counselors in 2004, out of misplaced bravado I said nothing about the dogs. Now I see how bound they were to the fabric of that search and the horrors they represented, but talking about them then seemed like an indulgence. Dogs were not what we'd been searching for. Grief might have marked me soft. I told myself I had written in my journal about it, had walked it off, and, a faithful woman, had asked God for some kind of wisdom that would explain how — and why — He would let those sparrows fall to the ground apart from the Father. I never got an answer, but hey, I'd taken the shot.

So I went back to work. I went out on other searches. I got on with things. That's been my way since childhood: When you're unhappy, get on with it; a 7 percent solution of self-preservation and denial.

Anxiety disorder began for me as a series of small nuisances months later. I started to have issues with music. An avid listener across genres, I was first unable to listen to orchestral pieces without feeling my spirits plummet. A heart-sink, I called it, and the sensation was as sickening as an elevator's drop from cut cable. *Weird,* I thought; *a mood,* I rationalized, but orchestra music was out. Rap from passing cars felt like a beating. Jazz was too damn glib, and pop music from any age seemed trite. Walking to a meeting in a downtown hotel, I overheard a lounge pianist selling Glenn Miller's "Pennsylvania 6-5000" and was furious.

That rootless anger was the next sign that something was wrong, and it attached itself randomly to little things — TV commercials, junk mail, and slow gas pumps. I'm normally a cheerful person, but sometimes then I felt so angry I thought my ears would explode, blown outward from the frustration of something simple — say, a dry-cleaning tag I couldn't remove. I have always liked to walk, and the urge to get on with things agitated me to walk even more. The madder I got, the more anxious I got, the more I walked, and in 2004, my weight began to drop.

A sudden crime wave in my neighborhood added to the problem. Burglaries. Robberies. Car theft. Rape. Bad news crackled through the area. Every incident seemed to bring trouble closer to home. A passing neighbor with her own dog liked to gossip about who'd been hit most recently. She had all the details: who'd walked in the front door to find the back door kicked ajar, whose gate was left open, whose pets were gone. She was a dramatic person given to exclamation. *"Busted out the back window in broad daylight!"* she'd say. *"Bam"* — making a Tae Bo jab with her fist — *"just walked right on in!"* My neighbor brought the bad word fast, and she tipped me to an interactive crime map on our city's home page. A person could keep up with what problems were happening where. *"Look at that crime map and see how safe we all aren't!"* she said.

Normally, I would have classified the woman as a local character. But now I was broody, so I checked out the map. Amazing. Five clicks, and a whole ugly local world opened up. The map had icons for assault, burglary, armed robbery, car theft, and homicide. Want to see all the area murders this year? Just check the appropriate box and click. Interested in stolen vehicles? Then deselect the little red icon of the akimbo man in bell bottoms (murder) and click the little blue car speeding away (auto theft). And if you reviewed all the crime in your neighborhood month by month, you could easily learn that a neighbor five doors away had had a burglary with garage-door access at two in the afternoon on the second Wednesday in June.

The second Wednesday in June? *Where was I at two o'clock that Wednesday? I was sitting right here.* Probably looking at the crime map.

I knew I was visiting the website too often — more often than it was updated — and that I stared at the same crimes every morning. But for me, the crime map had become as addictive as e-mail: What if something showed up on the map that hadn't been there last time I checked? What if it was on my street? What if it was next door? I didn't know enough about obsessive-compulsive behavior to recognize the symptomatic gnat cloud of what-ifs that was beginning to drive my impulses. But vaguely wondering if recent search-and-rescue callouts might have done me some kind of damage, I began to keep a journal of my visits to the crime map, hoping to sense what I was afraid of, or if I was afraid at all. Those days I was alternately angry, worried, and hyperactive. And skinny. Something was obviously bothering me, but what?

A morning run gave me part of the answer. Huffing through the neighborhood just after dawn one midsummer day, I passed a telephone pole swaddled in lost-pet flyers. I always stopped to look at these — no new habit there — but one particularly caught me. Beneath a picture of two tiny red dachshunds, it read:

Missing
Taco and Salsa
Lost during home burglary June 28th
We are heartbroken.
Reward for their return. Keep everything else.
No questions asked.

*Please.*

Standing there on the corner and sweating from the day's forecast heat, I felt cold. And I wondered if maybe I knew about this event already, if this family's break-in had appeared on the crime map with the little black mask for home burglary. I looked at the photo of the dogs, one dark as a kidney bean and the other the color of chili flakes, their heads tilting thoughtfully to the click of the camera. Were these dogs stolen, or did they simply walk out a door the burglars left open? For investigating officers, Taco and Salsa would be considered just part of the aggregate loss. Two pets with no street experience between them and high-traffic areas just a block away. Stolen or strayed, the prospects weren't good. (I could hear my neighbor in imagination: *And you know they're taking pet dogs as bait for dogfighting rings!* And then I realized I was hearing my own fears.)

*We are heartbroken. Keep everything else. Please.* I reached up to secure the flyer where the wind had begun to pull it apart.

So now that I was perpetually cranky and too well informed about crime in the neighborhood, it was a very short step to obsess over locked doors. I never worried about the house when I was in it. I never worried about anything anywhere else. But I was suddenly afraid that something would happen to my own dogs when I was gone. I began a meticulous leaving procedure. I would carefully put the small dogs behind two internal doors and Puzzle, my new golden retriever search-and-rescue partner, in her crate behind another door. Surely, I thought, burglars would leave the crated puppy where she was and avoid the big bathroom where a rabble of Pomeranians yapped and slavered. Burglars would strip the house of valuables and leave the dogs alone.

*Or would they?*

I got to feeling a little desperate. The dogs were all that mattered to me. There were wry moments when I wanted to paint a big sign for the living room that read:

DEAR BURGLAR
THE DOGS ARE ALL I CARE ABOUT
TAKE WHAT YOU WANT OTHERWISE
(THE TV IS OLD, BUT THE LAPTOP IS NEW!)

I imagined it like an old-school, hand-drawn grocery-store sign. With spangles.

Then dognappings for ransom started to occur half a mile from where I lived, and I was really screwed.

It got more and more difficult to leave the house. I would lock the door, take three steps away from it, turn around to double-check that it was locked, take three steps away again, turn around to check that it was locked again, and . . . you get the idea. Over and over, my thoughts raced ahead the moment I touched the doorknob, so when I got a few steps away, I couldn't be sure I'd really checked it. Even while I was caught in the loop of this, I knew it wasn't healthy. I'd scold myself — aloud — for the ridiculous behavior, then turn back to the doorknob a tenth time. Maybe the door really had come unlocked! Somehow I managed to get out of the house, but the compulsion was so profound that I started leaving for work twenty minutes earlier than I needed to, building in time for doorknob spasms if they should occur. (There was no rational logic here. I weirdly rejoiced in rotten weather, because surely burglars didn't break into houses in the rain?)

I needed help. I knew I needed help and didn't seek it, which wasn't wise. And it went on too long, that sense of loss and agitation and the compulsive lock-checking. It might have gone on longer, but, ironically, as I obsessed about keeping my dogs safe, one of them saved me.

Pretty Puzzle, the search-and-rescue puppy in training, was at her least patient and most willful in those days. I was a novice handler. Her obedience training demanded that we wrestle each other up the street twice daily on her walks, leash wars that pitted her youth and strength against every training technique I thought I knew. To walk the dog, I had to leave the house. Leaving the house with Puzzle required a whole series of obedience commands she deeply resented, and leaving the house with Puzzle while checking and rechecking the locked front door was a maneuver worthy of the Keystone Kops.

If anything productive came from that period, it was that, with all the lock-twiddling, I had to give Puz the Sit and Wait commands at least half a dozen times before we ever left. So she learned both pretty quickly. It went something like this:

"Puzzle, sit. Puzzle, wait!"

Lock the door.

Three steps forward.

Moment of panic.

"Puzzle, sit. Puzzle, wait!"

Three steps back.

Check the door.

Three steps forward.

Moment of panic.

"Puzzle, wait!"

Three steps back.

Check the door.

Moment of panic.

"Puzzle, wait!"

And so on.

This totally frustrated the dog. Initially she strained at the lead, refused the Sit, fought the Wait, and sawed through the garden on the end of the leash, trampling pansies and stomping columbine. Eventually she'd sit on command for a moment, then pop up like a jack-in-the-box, fight the Wait, and slice through the garden again. After a time, Puzzle sat and waited in a posture of great drama, her fuzzy backside on the ground but every muscle tensed forward at the end of the lead. And she'd sigh, big sighs that shook a little dog, like she was being good But It Was Time to Go Already. Youthful, cheerful, high-drive Puzzle was the very model of let's-get-on-with-it, the antithesis of my unspoken fears.

Puzzle badgered me free of that excessive caution. With her, I learned to leave the house more easily. I had to. I'd lock the door once, say "Locked! *Locked!*" aloud, and off we'd go. After a few months, I was able to leave cleanly with or without her — I was less anxious, more assured. In time, I learned to lock the door once and leave without a glance back. Was this a natural passage through the condition or did Puzzle's bright distraction ground me? I'll never really know. But I give her credit for the nudge toward healing, which began the moment I paid attention and stopped pissing off my dog.

# 3

THREE YEARS LATER, I am running after a full-grown golden retriever, my jog to Puzzle's easy canter as we thrash out of a search-and-rescue training sector at the edge of unfamiliar wilderness. It's a beautiful morning after a rain shower has washed the world clean — the air crisp, the kind of day that makes you feel like you could run forever. Puzzle and I are muddy and happy and unkempt, covered in wet leaves. Today's volunteer missing person thought to make things difficult. She cleverly hid in a tangled ravine where air currents could be tricky, but Puzzle found her in the way a search dog will, head up, *oingy-boing*ing through the brush as easily as we humans could if the victim sent up a column of thick, pink smoke.

Having left our volunteer behind for another dog-and-handler team, we stop to radio in sector finished and share a bottle of water. I flex my right foot, tingling in my boot. Puzzle and I have come out of the search area opposite where we entered, and because the sector we've been working is not square, or even rectangular, the route back through curving bands of woods will take some unfamiliar turns. I could pull out the GPS and tap a Go To command for the home-base coordinates, but I don't.

Next to searching, Puzzle most enjoys orientation work. "Take Us Back, Puz," I say to her, a command she understands. Depending on the circumstance, the *back* in *Take Us Back* means "back to base," "back to the car," or "back to where we started." Sometimes *Take Us Back* means going back the way we came. Sometimes *Take Us Back* means "Find the place we began but get there by a new route." This time we can't go out the way we came in, since it would disrupt the incoming dog team, so I'll let her make a new way back to base. I like to watch Puzzle pathfinding. I've come to trust her. Head up, tail waving, even though she's off-lead, she'll stay in my sightline. Puzzle's a field dog born to be a partner. A happy, independent creature that loves her work, she's glad to have this say in things and to show me the way.

It has been years since Puzzle's insistence that I walk away from doors I had already locked, and that compulsion seems so foreign now that it's difficult to believe it was ever my own. My young golden had made what mental health professionals might call a "behavioral intervention." It sounds brilliant and purposeful on the dog's part, but I'd call that intervention more a happy accident — the collision of her early disobedience and my willingness to give up my own *stuff* in order to train her. Whatever the motivation, the result was good. These are healthier days. I'm not only less anxious but also maybe a little wiser about myself — and I am curious and challenged by Puzzle, emotions that for me sit so close to happiness that I cannot tell them apart.

Can a dog really lead someone out of his own despair? If someone asked me now, I'd have to say yes. But much depends on the dog, and much depends on the human who follows him.

Puzzle's cheerful, "boldly go" nature still influences me in other ways. After one break-in too many in my old neighborhood and after years of thinking about leaving Dallas, I have decided not to sit and wait for a second round of fearfulness. I am ready to pack up dogs, cats, beds, and end tables. I'm ready to go. It's not a move made in any kind of panic. I'm simply ready to live somewhere else.

Flipping randomly through real estate listings outside the Dallas-Ft. Worth Metroplex, I find a historic property for sale that I recognize. I've known and loved that house for almost two decades, have watched it change hands several times. Wistful with each new restoration, I've imagined myself living in the modest Victorian cottage on the fringes of a once small prairie town. Boldly go — I put in an offer within forty-eight hours and move from the big city to my new/old house within a month, leaving behind the Dallas neighborhood, the crime map, and the cloud of worry that had once surrounded both.

That move brings air into everything. My dogs are excited — new view! New smells! New neighbors passing — *some of them with strange dogs!* I am happy too. In this place, old points of tension are no longer issues. Now a door is just a door. I have to lock it only once.

Between experiencing search dog Puzzle's lucky intervention and seeing assistance dog Haska's gentle skill beside her partner a few years later, perhaps it was inevitable that I'd become interested in dogs that serve the human mind. Search work has led me partway there already. Mental illness is at the heart of so many of our call-outs: missing persons who wander off due to dementia; who flee because of impulses born of autism, delusion, or anxiety; or who become despondent and disappear, burdened by depression they cannot escape. Any disappearance may be steeped in psychological subtext. Our on-scene family interviews often cover that uneasy ground, with questions that explore the missing person's recent state of mind. Obsessions, worries, memories relived, irrational fears, consuming grief, and expressed forms of guilt all come forward in many search situations, and often I enter a sector considering the terrain not only as I see it but also as the victim might have seen it, wondering what she might have been seeking there.

In fact, I had troubled victims in mind when I first began looking for a puppy to become my search partner. I wanted a dog whose face was especially open, friendly, easy to read, and soft. Before Puzzle was even born, I had decided on a golden retriever, thinking forward to the fragile missing persons she might serve. Puzzle is a search dog; she's not trained to intervene in psychiatric conditions, but she *is* trained to make her finds gently and not add to human distress. It's Puzzle's adult nature to be calm, to be kind.

These are qualities I also remember in Haska.

Puzzle and Haska serve at opposite points of a line I imagine: on one end, the search dog trained to find a missing human in crisis, and on the other, the psychiatric service dog whose work may help prevent that crisis. Despite the differences in the work they do, these dogs clearly share some attributes: intelligence, engagement, confidence, self-discipline, commitment, and — from what I know of Puzzle and remember of Haska — joy in their work. Psychiatric service dogs, emotional support dogs, therapy dogs — these are the dogs trained to answer human hurt. Similar to one another in some ways, markedly different in others, they tend all kinds of wounds. Just as I was drawn to partner a search dog more than a decade ago, I'm now spurred by the possibilities of service dogs too.

It is a difficult concept for many, that an assistance dog might recognize, for instance, the first moments of someone's panic attack and be able to intervene before it escalates, but to me it makes sense. If a search K9 can be trained to enter a crisis environment, make sound decisions, locate human scent, and communicate information to a handler, why *can't* an assistance dog be trained to recognize psychological events — by sight, by hearing, by scent — make a choice, and tell what he knows, in his dog way, to the human he serves? Dogs have been trained to serve partners with seizure conditions or potentially dangerous changes in blood sugar and have been doing these kinds of things for some time. Why not, then, train dogs to assist their partners with trauma-based flashbacks, wandering impulses, repetitive behaviors, and the like?

Plenty of people seem to agree. Working-dog experts, veterans' advocates, psychiatric-health consumers, and counseling professionals in the field are actively engaged in the psychiatric service dog partnership and are writing thoughtfully about it. Dr. Joan Esnayra, founder of the Psychiatric Service Dog Society, has spent more than a decade advancing the cause, publishing, in the *Journal of Psychiatric Services*, the first clinical case study involving the use of a psychiatric service dog. From that initial period of deliberation on the therapeutic potential of psych dog assistance, widespread partnerships have emerged — good dogs who intervene in human depression, anxiety, PTSD, panic attacks, bipolar disorder, and obsessive-compulsive disorders, among other conditions.

There are other needs of the human mind, and dogs that serve those too. Emotional support animals (ESAs) offer their partners steadfast companionship, usually in the home, and are not necessarily trained to perform specific service tasks. Though with medical documentation they are allowed to travel with their partners and to live in no-pet housing, emotional support dogs are not allowed in public spaces the way assistance dogs are. This difference sometimes tempts critics to write ESAs off as mere pets, but many emotional support dogs are something more than that. Through affectionate interaction, emotional support dogs may recognize and help lift their partners out of chronic depression; they alleviate loneliness. Often such dogs innately sense when to stay close, lessening a partner's anxiety. There are emotional support dogs that help their agoraphobic partners find ways to step out of their own houses. One ESA partner tells me that when laughter seems as out of reach as the moon, somehow her dog finds a way to make her do it. That's a common theme I hear from other partners: emotional support dogs often know how — and when — to be clowns.

Different from service dogs and emotional support dogs are therapy dogs, which serve a wide number of people. *Therapy dog* is a term in flux. It has evolved to include dogs used in actual therapeutic counseling scenarios and dogs who provide animal-assisted activities, engaging with the public at the sides of their trained handlers. Therapy dogs are frequent visitors to schools, nursing homes, hospitals, and other care facilities. Specially trained therapy dogs appear at recovery sites post-disaster. These dogs bring their own brand of good cheer to people in need, whether they are frail from illness, at risk in school, struggling to testify post-trauma, or coming to terms with the loss of home and family.

As a search K9 handler, I've seen how much good trained therapy dogs can do in crisis. The affected public turns to them. Emergency responders need them too. In the absence of therapy dogs, people turn to the exhausted search dogs after a disaster, and because the dogs are kind and their handlers compassionate, it can be difficult to pull the search dogs away from a grieving community so the animals can rest. I've come to appreciate the disaster-trained therapy dogs and their partners who serve onsite beside relief agencies. They perform no small service. People gravitate to these dogs; they stand in line to pet them or stand in circles around them, sometimes simply extending their hands to dog noses in greeting, as if warming themselves before a fire.

There is something special that gifted dogs can give to humans in distress. What is it beyond that solid core of good-dog presence? Perhaps it's the service of deep, wordless affection, free of human judgment and human platitudes. Guide Dogs for the Mind, the Psychiatric Service Dog Society calls these canine allies. "Godsends," says a SAR colleague who remembers therapy dogs among the grief-stricken in recovery areas after 9/11. That broad enthusiasm is motivating, but the more intimate stories are sobering. Service to humans in psychological crisis is no easier — and no rosier — than search-and-rescue. Stable, intuitive, compassionate, *strong:* at the heart of psychiatric service are some very rare dogs, so suited to this work they would choose it.

# 4

I BELIEVE THERE ARE dogs full of good intentions that want to serve the people they love but don't know how. Certainly I've seen it often enough: the dog that rushes forward when his human trips and falls, the dog that lies close to a family member who's doubled over in sadness, the dog panting anxiously over a distress he shares but cannot understand. I think it must be difficult for these empathic dog souls, bound to us as they are through centuries of companionship. Like many of us, these dogs have an urge to comfort, perhaps even to fix. Like many of us, some dogs seem helplessly conflicted; they seem to grieve when whatever they have to offer isn't enough. But how they try — and often keep trying.

What is that caring about? Is it love? Plenty of dog enthusiasts say yes — dogs are capable of love in dog terms. Others scoff at the idea of canine love, suggesting that all that loyalty is really about a deeply ingrained sense of self-preservation (*When a human prospers, I do too; when a human suffers, I may not get fed*). Others might say it's a compassion born of a long history beside humans: we and the dogs have forged a common language; a kindred, symbiotic spirit. Still others might claim that the sight, smell, and sound of human grief arouses a dog's ancient stirrings — whether it's to protect whimpering young or inspect a wounded beast for dinner. The answer may be any, all, or none of these. What we do have is evidence of compassion. Many dog owners have stories of their dogs' obvious attempts to comfort them. Quite a few owners manage to shoot videos of these for YouTube.

Smokey and Misty, rescued Pomeranians now a part of my family, are two such dogs. Both came to me after the death of their owner from cancer, a hard journey they chose to take as far as they could beside her.

While some friends fled, and family struggled, and there were those who couldn't bear to visit Erin as the disease progressed, these dogs elected to stay close to her when the crisis — physical, emotional, psychological — was at its worst.

Misty was five when she arrived at my house; Smokey had just turned three. The two are polar opposites in appearance and temperament. Misty, whose back legs are severely crippled, is petite, mellow, and quite beautiful — black, tan, and white, with delicate front paws that make her look like she's wearing gloves. Misty is a sunny dog. She's self-possessed and confident, except when she's nervous about being stomped by a larger housemate. She is protective of her frailties, and while she doesn't seem to suffer pain from the condition, she doesn't mind growling at other dogs that tumble wildly too close to her space, particularly when she's lying down.

Her adopted sibling Smokey is said to be a fifty-fifty mix of Pomeranian and Chihuahua. He must have been parented by a big Pomeranian or a big Chihuahua — perhaps a genetic throwback to the days when Poms were bigger. Smokey's a large boy, eighteen pounds, and he's a little oblivious, or passive-aggressive, or both. He is just the kind of dog that would body-slam Misty, and he does, now and again. He's also high-strung and reactive, with large, perceptive eyes. He is smart, smart, smart. Smokey watches, and Smokey listens. He's aware when humans speak his name, and he also seems to be aware of human intentions toward him when his name isn't spoken. It's not uncommon for me to decide that it's time to clip his nails or reach, with elaborately casual movements, for the shampoo bottle and find that he's disappeared under the bed and is tucked against the farthest wall.

In the dogs' first home, with Erin, Smokey was the favorite. For reasons I was never able to really understand, Misty was the cared-for-but-somehow-dismissed dog of that family. Perhaps it was because Misty brought trouble early — as a puppy, she once ate a bellyful of carpet tacks left carelessly behind by contractors, ate them before her mistress ever saw them underneath the couch, requiring a thirty-five-hundred-dollar surgery that nearly killed Misty. She returned home frail, withdrawn, and very much desiring to sleep on the pillow beside her mistress, and though Erin and Misty were physically close every night, some emotional connection between them was broken, if it had ever been there in the first place. Erin said that while she was fond of Misty, the little Pom was not all that bright. Erin kept her affection at arm's length. She was seeking another dog, a different dog, the kind of "heart dog" that Misty would never be.

Erin found that kind of relationship when she brought home puppy Smokey, whose bright eyes, upstanding ears, and instant devotion were an immediate hit. The two clicked where Erin and Misty never had. Erin was honest about it: Smokey became the favored, overindulged dog, and Misty was the watcher from a cushion half a room away. At night, Misty still lay on the pillow she'd always favored on Erin's bed, but Smokey slept pressed close to his mistress, his chin cupped in the palm of her hand. Misty was marginalized; Smokey got away with outrageous misbehavior.

And then their joint world fell apart. Erin had been feeling unwell through the summer of 2005, and in early fall symptoms progressed to the point that she began to do some research. The implications of what she found were ugly: colon cancer. She sought medical attention and returned home from the doctor angry with a sense of dismissal, feeling her concerns had been trivialized. "Get off WebMD — and see a therapist," she'd been told, but when symptoms worsened and she went for a second opinion, a round of tests and exploratory surgery proved her worst fears true. Erin's colon cancer was not only present, but far advanced. She was told that with chemotherapy, she might have a year. Without it, six months.

Chemo or no chemo was a decision she'd need to make soon. Erin had dreams those first nights, she said, of being alone on an unlit planet with the white arc of a rock coming toward her. Sometimes she flailed at it and the impact took her down. Sometimes she missed and felt the world fall away. But she always knew it was coming, and she would scream, whatever the outcome, a cry full enough that Smokey washed her face to wake her, and Misty would snuggle tight against her head.

Erin opted not to get chemo, and in the weeks following she began to *get her things in order,* as recommended by her follow-up counselors. It is a situation medical professionals must see often, but it was all new to me, a friend and bystander, who could help in only small ways as Erin carried the weight of a certain future before her. She was caring for a fragile mother also in poor health; she had a job, a mortgage, and two dogs whose fates were now in jeopardy.

The house could be sold. Her mother would live with her sister. But what to do with the dogs? If it weren't for her crippled back legs, Erin thought, pretty Misty might be adopted quickly. Smokey, she feared, would end up in a pound and euthanized. Erin needed to find safe harbor for both of them, but most especially for Smokey, whose hyperreactivity put strangers off but whose steadfast devotion broke her heart.

*I have bad news that will not end well for me. Will you help me find homes for my dogs?* she wrote to me just days after her surgery. I could sense the panic behind every minute lost. I already had a houseful of dogs — my own and a few fosters — and she was two hundred miles away, but I agreed immediately, thinking how hard a thing it would be to face your own death with the added despair of knowing you'd failed to protect beloved animals. We agreed that I would foster her dogs when the time came, get the word out, and thoroughly interview potential adopters, but as Erin's illness progressed, her question changed from *Will you help me find homes for my dogs?* to *Will you take my dogs?* Even though I assured her that I would shelter Smokey and Misty and find homes for them with the same care I would find homes for my own, that was not enough. Erin redirected the energy of her grief into persuading me. She needed to imagine them with a home. She needed to know they'd be in mine. She begged outright.

I said yes, but she wasn't completely satisfied. *Are you sure?* she pressed. *Are you sure?* I told her my only reservation was how Smokey would adjust, moving from the center of her universe to a noisy household full of other dogs. If I could find the right home for him, might he not be happier if he was someone's only dog? Erin was uncertain. As her condition worsened, she seemed less sure that I would find the right place for Smokey. She had nightmares of Smokey ending up in a home where he'd be abused. She wanted to die knowing absolutely where he would be.

I promised her I'd keep him, and I meant it.

Sometimes my assurances seemed to comfort Erin, and sometimes they did not, and I began to see the role all of this played in her struggle to come to terms with approaching death. She needed a project, and so every visit I made was a chance for Smokey to audition. Though Erin was frail, she'd take Smokey to the groomer before I came. She bought him doggy breath mints. She spent a lot of time selling me on his merits: he was a good dog, the best dog ever, she'd say, cuddling him and shooting an anxious look in my direction when Smokey barked at the mailman, the microwave, *Jeopardy!*'s jump to commercial.

Once he barked over a single leaf falling from a tree.

"A watchdog!" Erin praised.

"Genius!" I agreed. (Oy.)

One day as I played with the dogs to amuse her, I threw a tiny stuffed giraffe, and Smokey retrieved it. Erin laughed aloud for the first time since her diagnosis, clapping her hands, making her mother jump, surprising all of us. She held her hand out for the toy.

"Fetch!" She threw the toy, and back he came with it again. She had never tried fetch with him before. Once might have been a fluke, but twice was brilliance. "Smokey fetches!" She shot me a proud look. "This dog," she said. "What's not to love?"

Smokey's timing could not have been better. Erin relaxed, certain I would now take her dog, because he fetched. Smokey, in turn, was thrilled with her reaction. He fetched, and fetched, and fetched again, joyful with Erin's pleasure and my applause. From her pillow at the edge of the room, Misty watched.

As best she could, Erin tried to seal the deal by teaching Smokey other attractive tricks. Smokey was quickly good at catch; he enjoyed the lie-down-roll-over routine, and he was glad to show off a tail-chasing maneuver Erin called Pong, like the old video game. In Pong, Smokey chased his tail, collided with furniture, bounced off the furniture, and then — still spinning — bounced into something else. "Pong!" Erin would say, and Smokey, self-winding, would dervish-whirl across the room — couch to table, table to rocking chair, rocking chair to, occasionally, Misty, who first skittered when she saw him coming and then later wanted to get in on the action. She began to intercept Smokey, scrambling to where he was like a soccer goalie and shrieking *Yaagh!* into his ear so loudly it would redirect his course. For a disabled dog, Misty had game.

If there was anything bright to be found in the last months of her life, Erin found it with her dogs. Sometimes when she sat weeping in front of the television, Smokey would start up Pong on his own. He had learned she was her happiest playing with him.

Too soon, my friend answered the phone less often, and, her family said, left her bed more rarely. Hospice caregivers began to come in daily, and one late-autumn day, Erin called, her voice faint and sluggish, to say that I should be prepared to take the dogs sooner rather than later. There were new problems. Her condition had deteriorated, and Smokey, witness to his person's nightmares, falls, and meds-induced confusion, was growing too protective. He had begun to station himself on the bed and growl — even snap — when caregivers attempted to help Erin, once blocking the door when paramedics were called to assist. Smokey had grown frantic with shared misery. He was anxious and overwrought with impulses to protect her and cheer her. During her worst moments, Erin said, he'd begun to bring her his fetch toy, the only fix he knew. She'd awakened more than once to find he'd put it in her hand. Misty had planted herself on the pillow next to Erin and seemed unwilling to move, scratching Erin's pillow whenever she had nightmares.

"This is no good for any of us," Erin said. "You can take Misty at any time, but Smokey — how can I bear to be without him?"

One thing was certain: Erin was too weak to make decisions. I spoke to her family. We strategized how to keep Smokey with her as long as possible and how to keep him away from the caregivers when they came.

That late-autumn day was the last time I ever talked to her. Erin slipped into a coma within a week and died three days after that, Misty on her pillow, and Smokey by her side. The dogs lay beside her for two hours after her death, and then they seemed to recognize it was over. When it was time for Erin to be taken from the house, the family was prepared for trouble from both of them, but Misty crawled into a bathroom and Smokey, subdued and bewildered, let Erin go — all his protective instincts gone. Both dogs alternately hid beneath the bed or wandered aimlessly in silence in the days prior to coming to me. Misty chewed a raw spot on her foot. Smokey's fetch toy disappeared.

I believe that, in their dog way, which is no less soulful than ours, many dogs understand death, and I believe that they grieve. I also believe that both Smokey and Misty attempted to give care to Erin in that dark period. Other adults in the house had long been responsible for their feeding, their walks, their vet trips and grooming; their own dog needs were being met. But though they were not formally trained for service and had not learned specific tasks to address Erin's nightmares, panic attacks, and depression, the two chose to stay close, offering comfort in the only way they knew how. The desire to help her was there. Smokey's protectiveness and desperate offers to fetch also spoke of an urge to make things right. Misty's pillow-scratching in response to Erin's nightmares — was she trying to wake her? — seemed an unanswered reaching-out. It must have been difficult for them, bound in love and a confusion of impulses, sensing their shared world slip away.

Soon they would enter a new one. Survivors Smokey and Misty would come home to the rodeo of cats and dogs that was my house. It was hardly a peaceful transition. Their confusion was as clear as their great sense of loss. Who were the animals and I to them? What would become of them here? When I think of Erin's dogs in this interim, I remember most their silent watching, huddled together and apart from us, their rivalries spent.

## 5

ONE THING IS CERTAIN. Dogs that serve the human mind are a lot better known now than they were a decade ago. There have been some good books written; there's been some media attention. Therapy dogs have shown up center stage in feature stories set in crisis centers and health-care facilities. The emerging use of psych assistance dogs with U.S. soldiers returning from the Middle East has increased public support. The stories are riveting. We've seen dogs recall fading seniors to happiness and connect autistic children to the world around them.

The attention has also rapidly increased demand. Aware now of the dogs' therapeutic potential, more care facilities are utilizing therapy dogs or placing companion dogs in-house, and more mental health professionals are recommending psychiatric assistance dogs or emotional support dogs to their patients.

Many service dog organizations have stepped up to meet that need, expanding their services to include psychiatric assistance dogs. (Some also offer trained therapy dogs to family partners who would like to serve the public.) It's a thoughtful process: for assistance dogs, an organization typically matches the dog to his potential partner after general obedience and service training is completed, and then it trains that dog to meet the handler's individual needs. Such skilled assistance comes at a price. Raising, training, and providing excellent care for assistance dog candidates is an expensive proposition for the service dog organization. Assistance dogs can cost thousands, and a prospective handler may have to fundraise for years in order to acquire one.

There are other options. For some, an owner-trained service dog — an animal that has the same legal protections as a program dog from an organization — is a viable option. It's not a choice that's right for every would-be handler: a young, untrained dog, however bright and sweet, may be too much for a novice partner with conditions that interfere with successful foundation training. But for others, building the early relationship with the service dog in training (SDIT) and developing a dialogue specific to the handler's need has real benefit. The Psychiatric Service Dog Society is a staunch supporter of the owner-trained partnership when it's appropriate, but the group's literature strongly urges handlers to seek the assistance of a professional in the thoughtful, responsible choice of dog and with the training involved. Feedback from an objective third party is good.

There are critics of owner-training, of course, those who find it difficult to understand how a mentally ill person can reasonably teach a dog to sit and stay, let alone intervene in his own condition. (One handler told me, "Too many people assume mental illness means your head is always about to explode.") But such concerns aside, for many this is a form of cognitive behavioral therapy: a handler has to analyze his own psychiatric events, identify symptoms, train the dog for tasks she can do to intercede, and, perhaps most difficult, train himself to pay attention to the dog. The process creates the language between the dog and handler. Challenging, maddening, rewarding — many handlers find they learn as much about themselves during the process as they do about their dogs. Program dogs versus owner-trained dogs is an ongoing debate, with advocates and detractors on both sides. I'm always wary of the notion that there's only one way to do anything — particularly with dogs — and while I've seen many fine examples of program-dog partnerships, I understand the argument for owner-training too.

But caution to anyone who thinks that the owner-trained option is a cheap and easy way to sidestep the formalities of acquiring a program dog. There is much more to it than the I'll-feel-better-if-I-have-a-dog-with-me-so-I'll-just-put-a-vest-on-my-dog-and-go approach. For all its support for owner-training, the Psychiatric Service Dog Society is straightforward: This process is rigorous. An owner-trained dog and his handler should meet the same high standards that a program-trained dog and his handler do. "Higher, even," Joan Esnayra commented to me on the phone. In a world where invisible disabilities and the dogs that serve them are often suspect, an exemplary owner-trained partnership makes the best case for their existence.

Handlers tell me the healing power is in the dialogue. For them, psych dog partnerships make a difference other therapies cannot. German shepherd Neo is a good example. Neo serves a veteran who has not truly been able to leave the battlefield. Still caught in a moment of in-decision during an attack that cost the lives of civilians and fellow sol-diers, Neo's partner is left with panic attacks and chest pain, a door fixation and the perpetual taste of blood in his mouth. He was pretty wound up right after it happened, he says, but it all came on gang-busters a few years later, when his body began to fail. The idea that he might have to live with caretakers for the rest of his life made him so depressed he considered suicide. He's still young. He wants to live independently, but he needs an ally. He has that in Neo, a social, gentle, re-careered police K9. Neo is diligent, and he is stubborn. He's a dog that can insist when he has to. Sometimes his former soldier just locks up, and Neo is trained to recognize it. In those dark moments, without Neo to nudge him through doorways, his partner could not get from his bedroom to the kitchen, or go to work, or escape a fire.

Consider one partner — we'll call her Melissa — the long-term victim of a child predator. She is in her thirties now and has multiple anxiety disorders. Though Melissa knows a few of her triggers and avoids the ones she can, there is no way to predict some attacks. A specific scent, a certain time of day, a particular slant of light — she is in a public build-ing when she suddenly fears she's being followed. The sensation is ter-rifying and very real. Melissa takes an elevator up to anywhere, arrives on a floor she doesn't recognize, turns down a hallway, follows several corridors in a panic, constantly feeling the presence of an oppressor behind her. She finds a bathroom or a closet and hides. As the event evolves, she might creep from the closet with little sense of time and space. She wanders blindly, completely disoriented, with no idea where she is or how she got there. Walls bend and floors curve. She is dizzy. In some severe cases, Melissa may fall to the ground and have difficulty feeling her hands or feet.

A psych dog can preemptively help partners like Melissa distinguish what's real and not real. Melissa knows that her PTSD flashback episodes often involve the sensation of being followed, sometimes to the extent that she thinks she sees a figure behind her, so she can train her psych dog to indicate the position of strangers on command. The dog acts as a reality check. In that first uneasy moment in the empty atrium, if her dog does not indicate the presence of a stranger, Melissa has solid feedback: *There's no one here to follow me.* Or if he does: *He's indicating a woman on my left and now two men walking out the far door to my right, but he doesn't acknowledge the person I think I see.*

If Melissa has attempted to get away and is now lost, she can forestall the resulting panic by telling her trained dog "Backtrack," a scent-driven task that requires the dog to find and follow their trail back through the maze of hallways, down the elevator, and to the starting point. Now acting as a guide, Melissa's dog can steer her around hazards. If she attempts to wander onto a busy street, her dog can block her from doing so. Melissa can give the command "Brace," and the dog will stand sturdily on four legs as a support, to help her stabilize. If Melissa falls down, her dog can stand over her and bark for help.

It sounds a little like fantasy and a lot like Hollywood, but for some psych service dogs, this is business as usual. Partners train with their dogs for events they hope will never happen, but they train hard, expecting that they will. And in Melissa's dog's case, the service starts with learning basic tasks: acknowledge a stranger; find the door. Called upon to do these tasks and more — seek help, clear airways, redirect behaviors, urge housebound partners out, lead lost partners back home — psych assistance dogs must be extraordinary, and their partners no less so.

Experts suggest that those best suited for psych dog assistance are people who naturally like and trust dogs as well as understand their own disorders. These individuals have a good idea what triggers episodes; they know how their conditions present. In many cases, Joan Esnayra notes: "Psych dog partner candidates have often been there, done that, tried the shock treatments and the meds cocktails, and are willing to make a serious investment in their own wellness."

A certain amount of stamina is required of handlers; the partnership comes with significant obstacles. Living in the netherworld of invisible disabilities, psych service dogs and their partners encounter frequent discrimination. While the Americans with Disabilities Act protects the partners' rights in most public venues, the Air Carrier Access Act requires official documentation from psych dog handlers that no other service dog partners (including medical-response dog partners) have to provide. The prove-it moment, as one handler describes it — when a gatekeeper doubts you and the dog — is deeply personal, often uncomfortable, and sometimes adversarial.

In some places, legal protections for psych assistance dogs are clear; in other places, gray areas leave room for interpretation — and trouble. An Internet search brings up plenty of examples: Handlers interrogated at airline ticket counters, angry shop owners calling the police, disabled partners or their dogs threatened in front of a host of strangers. Dogs kicked. A Georgia handler struck by a restaurant manager. A nursing home in Lexington, North Carolina, that allows the presence of therapy dogs for its patients but refuses to let an employee's psych service dog on the premises. In one small Iowa town, officials banned a fully documented pit bull–mix service dog from living in the city limits at all.

One handler comments, "It is ironic that beside the dog trained to help you with an anxiety disorder, you can end up in situations that would make most anyone anxious."

Another handler notes: "And here's the kicker: It's a kind of trade. For all the good help a service dog gives, that same dog makes you visible. If for you a 'normal' life is about being able to be anonymous, good luck. You and your dog and your disability had better be prepared for stares and questions . . . and sometimes accusations. It's not all bad. A lot of the attention on the dog is supportive. But there's almost always a spotlight. It slows you down."

In addition to wanting this help and being willing to work for it, therapists, handlers, and trainers agree, human partners *must* be able to provide good care for their dogs. Assistance dogs have lives and identities and needs of their own. A good partner candidate recognizes the responsibility to provide a full life for his service companion, which includes a commitment to the dog's health and safety and allowing for downtime and off-duty play. Some handlers say they were surprised by the benefits of this dog side of the relationship — when a dog's high energy leads the pair to agility competition, for example, and the agoraphobic human with profound social anxieties now finds herself in front of an audience, running beside her joyful dog, by choice.

The benefits go both ways. The dogs attend to their handlers' conditions. The handlers become keenly aware that their dogs are individuals. Yes, they say, these are the dogs that make a whole life possible. These are the dogs that save us in the big moments and the small ones. But many handlers are also cheerfully honest: These are good dogs, but they are creatures with foibles too. There's the mobility dog that hides his ball in his partner's shoes; he'll bring the shoes on command, then — *Oops, how did my ball get here?* — offer a game of fetch. The dog trained to open a refrigerator who, as a youngster, occasionally sneaked himself late-night snacks. The dog that barks at butterflies. The dog that wants a lap of morning coffee. There's the landscape digger. The bed-hog dog. The dog that flirts with his own reflection when he sees himself in any kind of glass. That dog's handler showed me a cell-phone capture of a dog coyly tilting his head to himself in a hotel mirror. "I could take this dog to the White House," he jokes, "but I could never take him to Versailles."

How difficult is it to find a dog that's right for psychiatric service and, with expert help, owner-train that dog to a high standard? After months of research, I hope to find out. I've put out the word about my intentions, and through friends, professional organizations, and social networks, people have come forward to help. Many disciplines shape this work. I need input about all of them. Assistance dog partners; therapy-dog handlers; dog theorists, evaluators, and trainers; psychologists and therapists generously offer their expertise when I ask for it. Six handlers tell me their in-depth case histories. Many introduce me to their dogs. Better informed by their experiences, I hope to find a dog with a strong aptitude for assistance, to train that dog to reach the public-access standard and to perform a handful of typical psych-service tasks, and then — we'll see what happens.

I'd like to be a resource for others. If the chosen dog does very well, he could remain with me as a demonstration dog to help other new handlers training their own assistance dogs. There's also the possibility that someone with an urgent need might come forward, and, if the match is a good one, the dog could become that person's partner. And of course, if the chosen dog proves himself unsuited — too field-driven for assistance work, too anxious in crowds for therapy, or too unmotivated for either — then he'll have the opportunity to learn something else, or he can simply remain a loved family member, wherever his dog gifts lie.

There are other possibilities. I try to avoid thinking about this, but progressing kidney disease has already begun to cost me strength periodically and, on the very worst days, shows up as peripheral neuropathy — numbness in my hands and feet that affects walking and particularly moving up and down stairs. Though I'm not disabled, it's not difficult to see that in time, I might need an assistance dog to help with these issues. Not soon, I hope. I'm still working search-and-rescue. I still have more good days than bad. It's my nature to move quickly and alone.

But I use a walking stick on the very worst days at work. In the field, I've noticed that SAR colleagues are more often helping me to my feet when I have hidden for their dogs, clearly noticing something I would rather ignore. When I mention to my doctor that I am exploring work with service dogs, he looks at me thoughtfully and says, "Well, *that* may come in handy down the line." That I am planning to work with psychiatric assistance dogs doesn't faze him. He thinks mobility tasks are also in their skill set. "You'll be all kinds of balanced," he says.

# 6

THE GOLDEN RETRIEVER IS frustrated. We are sitting with friends at a restaurant that allows dogs on the patio, and Puzzle is in her good-dog Down/Stay beneath the table, her head resting on her paws. I don't think my search dog views this brilliant Sunday morning the way I do, but I can see her nose twitch and her brow furrow. I know she's working the light winds where she lies, enjoying all the scents that make up the cool air we call Texas Champagne when we're lucky enough to get it. Puzzle would much rather be in the field on a morning like this. She is not really a brunch kind of a dog, but here she is with me, being a brunch kind of a dog. She is very, very good and very, very bored. My friends don't notice it, but golden owners know the breed's great gift of communication, the power of that golden's deep sighs. I feel Puzzle's pointed huffs against my foot. We've been here for a while, and at first, every time I moved my chair, her head snapped up, like she was hopeful of parole. Now she's refusing to look at me at all.

Since I have a dog like Puzzle in the house, my friends wonder why I spend time looking for another dog to train for service. The whole find-and-train-a-rescued-dog initiative they understand (although it could take *seriously forever*, as one friend put it), but the lines between search and service are blurry for them. All it takes is a smart, friendly dog who'll learn to do something and do it reliably, right? They know Puzzle well, and it seems to them she's right on point to train as a service demo dog, or even as my own assistance dog, if I need one down the road. Another trick in her bag of tricks, they say. What could be a more natural add-on to her resumé?

They're right, at least in the broadest strokes. Puzzle is smart, friendly, and polite in public. She is eager to learn new tasks, and she is bonded with me. She's theoretically perfect. The fine points that don't make assistance work right for her are more difficult to explain. Though Puzzle's aptitude testing early on as a puppy (at six and ten weeks) and later as a young dog (about seven months old) showed all those good qualities of engagement, eagerness to work with a human, and stability we want in a working dog of any kind, early training with her also made it clear that she's a field dog at heart. Puzzle is strongly independent, and given a choice between a job that requires her to be at a partner's side on a leash all day and a job that allows her to blaze an off-lead trail to find a missing person with her human partner in follow, Puzzle would always choose the latter. Her affection for me is deep, but that affection formed in the field.

Sure, Puzzle has shown genuine concern for me the one or two times I've been hurt, and yes, I've realized recently she can identify the chemical changes created by my kidney condition. She gets a little more attentive, seems to know when I'm going to have a bad day before I do. While I have no doubt Puzzle could be trained to do assistance work, I am just as certain she shouldn't be. Puzzle has little desire to go everywhere I go if there isn't a search at the end of it, and I could ruin a good dog by forcing her in that direction. I'm looking for the dog that not only can do assistance work, but wants to.

That said, Puzzle's love of orientation — a fairly common psych-service task for dogs with a gift of nose — can teach me a great deal. She has already shown me her location skills in the search field, of course, where she has long been able to make her way back to incident command without much guidance from me. Though I can only guess how she does it, I believe she remembers the progression of scent and its changes as we move through an area. These are landmarks by nose, so for her, it's a matter of leading us from the edge of a cow pasture back to the smell of the creek, and then to the field where hay has been mown, and then to the abandoned gas station, and so on. The unique scent of each space, although too subtle for me, probably paints the world much more clearly to a search dog.

A few years ago, Puzzle and I were traveling a lot. I was already thinking forward to psych assistance dog orientation tasks, and it seemed like a good opportunity to explore her skills. We made a game of it. We'd check in to a hotel room, leave the hotel for a walk, and on return, I would say, "Where's the room, Puz?" and let her lead me there. *Where* is a word she already associates with searching, and it didn't take much for her to learn what *Where's the room?* meant. At first, I didn't ask until we got off the elevator on the right floor, and then I'd let her choose which corridor, which turning, and which door amid all the identical ones was the one to our room. On later travels, I'd ask her in the lobby, and invariably she'd lead me to either the elevator (did she identify it by our previous scent? Or did she remember it visually?) or a stairway. While Puzzle certainly can't choose the elevator button to push — she doesn't bark five times for the fifth floor, by any means — the few times we were in an elevator by ourselves, she could discriminate which floor we should get out on if I pressed three or four buttons and had the door open on multiple floors. She would stick her nose out and make a yup-or-nope decision. Once on the correct floor, she seemed to find it easy to choose the hallway and the correct room — even after going into and coming out of that room only once.

All of this must be strongly related to her search work. We talk about a dog's baseline scent — the scent landscape a dog paints when he enters a particular environment. Just as we identify visual landmarks in new places, the dogs seem to create their own scent-driven ones. (They may also be making visual connections, but if so, my dog, at least, isn't telling.)

That lifting of the nose and making a yes-or-no decision is also part of search. On the very fast clear-building search drills, a dog has to be able to stick his head into a room and, in just that tiny moment, decide whether there's human scent there or not. I think there's an essential relationship between that and Puzzle sticking her head out of the elevator to sample the smell of the corridor, only in this case, she was searching for the remnant scents of us.

One autumn, I was invited to give a presentation at a beautiful lakefront spa in Austin, Texas. Puzzle and I drove down from Dallas, arrived late at night, and received the key to our cabin. It was one in a row of about forty almost identical cabins along a short bluff. Even though we'd been driving a long while, the idyllic setting, flush against a lake where we could see white birds skimming low, was too tempting not to explore. We got to our room, unpacked, took a quiet walk, and went back in for the night.

The next morning Puzzle and I set out to walk again, and this time I decided that we'd take the cinder path that fronted all the cabins and go as far as it went west and then come back again and go as far as it went east. Then we'd take the staircase at the east end to get down to the lake. It was a lovely morning. Passersby were friendly. Many of the spa visitors were having coffee on the porches of their little cabins, and we lingered with some of them before continuing on. After descending the far stairs and exploring the lakefront for an hour or so, we headed back to the cabin, and I decided to take a central staircase that we'd never used up to the bluff's cinder path. I dropped Puzzle's lead at the base of the steps, took up my phone and started the video camera, then said, "Where's the room? Take Me Back, Puz."

We'd been working on this the way an assistance dog must do it, and Puzzle had learned that when given the command on-lead, there was to be no trailblazing — it was a signal to go slowly enough that a human could follow at a walk. She moved calmly upward.

We arrived at the top of the steps in the middle of the cinder path. This was the path we'd taken this morning, end to end, and we'd done a little bit of it last night too; certainly, our scent was scattered all along it, but there should have been at least a little more of the smell of us at our own cabin. I wondered. We had been in this place less than a day. We'd gone in and out of our cabin only twice. Since she was still working the orientation command, which cabin would Puzzle choose? And how would she distinguish between our cabin and the identical porches of those we had visited?

At the top of the steps that led to the cinder path, Puzzle took a step forward, paused, turned her head left, then right, then left again, trying to decide. "Work it out, Puz," I said, the words of encouragement we used on searches too, and a moment later she chose to head left, lowering her nose slightly and trotting along the path toward our cabin, dragging the lead. I followed behind with the camera. As we passed each cabin, she turned her nose in the direction of its doorway and then rejected it — just a flash of a scent check. This one? Nope. This one? Nope. Until she reached the steps to ours. She turned right, led me to the door, and looked back to beam at me. It was the same expression she made on some search finds, happy with the challenge and confident she had it right. She did.

I understand the process. I like to imagine the dogs' thoughts as they mark scent landmarks the way we would note visual ones (*I remember passing this convenience store and this house with the funny mailbox,* a human might think, only for a dog it might be: *I remember passing this splash of bleach on the sidewalk and this cabin with the faint scent of Tiger Balm, and, oh, two squirrels scuffled right about . . . here*). Still, the demands of scent discrimination in a place where every doorway seems, to me, identical; the ability to trace scent back to a starting point — it amazes me every time. Take Me Back was a command we'd continue to work. A search dog's talent, a service dog's task — it might be a skill Puzzle could share.

# 7

A 2009 NATIONAL PUBLIC RADIO story underscored what most of us who work with dogs already knew: While the homeless-animal population has long existed, the Great Recession has made things worse for American pets recently. In some areas, intakes have increased annually up to 400 percent. Surrendered dogs are often healthy pets that are up to date on their vaccinations and have been neutered or spayed but have to be given up because their owners have lost jobs, or homes, or both, and they need to find safe places for their dogs to go. It's a desperate choice made with loving intentions, but for these surrendered pets, too often there are not enough adoptive homes to go around. Rescuers have had to work even harder than usual, trying to maximize adoption events and social-media exposure to give more homeless pets some kind of hope.

I've often thought my next search-and-rescue dog would be a rescue. These are the dogs I'd like to work with in psych assistance and therapy too — the homeless ones, the last-chance dogs. I'm not the first to think of rescues for service, certainly; there are fine assistance dog organizations that have used dogs drawn from the rescue population for years. Many groups acquire pups from respected breeders or have developed breeding programs of their own in hopes of achieving a greater percentage of graduate success, but the assistance dog programs that use rescues have chosen a different option, adding the complication of unknown backgrounds and genetic histories in many of the dogs they assess. It's a fiddly proposition. Evaluating rescues for service is a difficult process involving a population of dogs in need and a responsibility to the people those dogs will serve — heroic and heart-rending work. I understand the difficulties, and I take them very seriously, but having seen rescued dogs successfully chosen, trained, and released to jobs they enjoy and new lives in which they thrive, I'd like to try. I want to find a good rescue candidate and train him for service.

Talented shelter dogs are not new to my friend Tom. He's an advocate for homeless dogs, and years ago, on a small scale, he assessed them in animal-control facilities for search-and-rescue and accelerant-, drug-, and explosive-detection work, with good results. He's also one of the best 360-degree thinkers I know. When I tell him I'm interested in working with rescued dogs to train them for psychiatric service, emotional support, or therapy, he has to consider it a moment.

He says I'll probably find some dogs that might be candidates for therapy work. But rescued dogs for *psychiatric service?* Tom remembers the dogs he encountered in pounds. Many of the ones he saved were high-energy dogs surrendered by overwhelmed owners who'd been unprepared for all that intensity or the hungry, homeless adolescent dogs that worried neighborhoods with their prey drive but that had not yet gone too feral. Such dogs needed strong handlers, but they were often perfect for detection work. The other dogs . . . the other dogs he saw were timid. Or reactive. Or unpredictable. Or sick. Or old. Some were aggressive. Some had bad hips. Some, he admits ruefully, were just not all that bright.

He says you need smart, trainable, self-disciplined dogs who like people. They should be solid — motivated, but calm too. But calm dogs make great pets. Even in a recession, how often do great dogs end up in shelters? Tom looks at his own rescued golden retriever, Camel, named for the cigarettes Tom gave up after a heart attack. Camel was surrendered to a shelter at a year old by an owner with a health condition who was too frail to give the big, amiable dog any exercise. Camel is, in fact, the very kind of shelter dog we are discussing. I don't say it. My friend doesn't say it, but he rolls his eyes and winks at me as we both overlook the great-dog obvious whose tail is *thump-thump*ing on the floor.

Tom warns me. Here's the deal about pound evaluations: Very few shelters have good spaces for evaluation. Very few shelters have the time to give you with multiple dogs. In some pounds, the conditions are so desperate that no dog will test well; they'll be too strung-out from the tension. While some shelters will allow dogs to be taken elsewhere for evaluation, plenty will not — most of them will not. Very, very few allow the option of returning a dog that ultimately can't do the work.

So at the shelter, a dog often has to be some kind of genius on the spot. Genius on the spot is asking a lot of the dog. Can an evaluator make wrong calls? You bet.

Out of nowhere, Tom says that finding and training rescues is worthy work. But he isn't sure that I'm the one to do it. He knows my history. He looks at me thoughtfully. "This doesn't take a hard heart," he says, "but it does take a resilient one. I wonder about you walking away from euthanasia-tagged dogs in cages. Susannah, do you really want to go there?"

That's all he says, and it's enough. No. I don't want to go there. All of the pets I had as a child and most of the ones I've gotten as an adult came from shelters — I gamely walked in and came out again a little later with a new furry family member in arms. But since that 2003 experience with those caged dogs left to die in the woods, I've found shelter visits almost unbearable. I don't want to even think about leaving more doomed dogs behind. Yet in this kind of work, that's inevitable. There will be dogs that don't evaluate well, and I'll have to walk away.

"Think about it," my friend says evenly, leaving the caution at that.

The next morning it's clear he spent the evening on the Internet. He mentions a handful of local dogs he found on Petfinder that might be good candidates. We are both saddened to read how many of these dogs are owner-surrendered. The reasons for the surrender — the owners' job losses or home losses among them — don't seem to matter; write-ups from volunteers suggest that even the smart, good, well-behaved surrendered dogs are getting little time. The kennel space and budget don't allow it. Shelter staff know these dogs aren't lost and that no one is going to come in and claim them. Tom muses about the dogs he had noted. If the write-ups by volunteers are even halfway true, with careful evaluation, some of these dogs might well be candidates for service. Not a lot, he says. Expect one in thirty. He notes a few on his list that might be better suited for ESAs or therapy.

Good has already come of this. My friend says he and his wife are going to check out a blind chow–Irish setter mix he found at a shelter not five miles from home, a bright red beast who's out of time and bound for euthanasia. "We'll bring him home. We'll call him Chili Dog,"

Tom says — another rescued dog named for something else my friend gave up.

On the other side of the country, a retiree walks the cages of animal shelters. She's an evaluator for dog rescue, a second-chance gal. Paula pulls dogs that need safe harbor while rescue volunteers try to find them permanent homes. She finds pets mostly, and she finds working-dog candidates every now and then. Paula's in these shelters often enough that the staff know her name and her blue canvas bag of tricks: the leashes and harnesses, clickers, treats, and toys, the rather horribly realistic rubber hands, made for testing dogs that might bite a human over food bowls. (Paula sometimes leaves the fake hands in the passenger seat of her car as a crime deterrent. They've been around the block a few times; they always stop people cold. The fingertip of a pinkie is missing from one of them, testimony to the jaw strength of a dog that did not pass the test.)

We are friends who have never met face to face. Paula and I became acquainted online about fifteen years ago through a dog-rescue transport group's Internet message board. In an environment of high stakes and higher emotion, the sky was always falling. Paula was all business on that board. I admired her frank, straightforward posts, her compassion coupled with practicality, and her refusal to take part in interpersonal dramas of any kind.

As with many others online, Paula uses a pseudonym. She is wary of anyone who seems too interested in when she'll be away from her house. She is away from her house a lot, on one road or another, pulling dogs from pounds when possible, taking them to rescue groups or new forever homes. She sometimes gives up holidays with her family to do it, a practice her grown daughter tolerates rather than admires, a practice that puts her one inch closer to being the Crazy Dog Lady in her neighborhood. With her drive to do a thing outside convention, Paula reminds me a lot of the search-and-rescue folk I know. We get along. We are similarly covered in dog hair. But in the area of evaluating dogs for service talent, I am the novice and Paula the expert. She has been going into shelters for years, and if anyone can make peace with the mix of hope and futility to be found there, she can.

Paula has evaluated enough of these dogs that she expects failures. She grieves the lost causes anyway, the ones who can't pass the test for service and, even worse, show themselves to be unadoptable even as pets. She says you have to learn how to grieve and get over it. There will be new dogs in equal jeopardy tomorrow. There are always more dogs than cages, and shelters move so damn fast. The young, stable dogs that come in without too much baggage earn a little more time.

Paula always keeps an eye out for the special dogs, dogs with intelligence and stamina, dogs who've had a less rough time of it or who, in the surprising depth of their resilience, have had a *worse* time but who make better choices beside humans. She is looking for dogs who are eager, responsive, and willing to learn, who might make working matches of any kind for partners who need them. She wears a brown jacket that doesn't show dirt and hides the dog hair that she rarely lint-rolls. The smell of pee and bleach has stopped making her nose run.

She says it would be easier if she didn't love dogs. She has dreams where every one of them takes the test and passes. But that is never true in real life. Paula takes all evaluations seriously. Pet evaluations are grounded in safety. Evaluations of potential working dogs are a life-and-death matter on both sides, really. Sometimes she has a day or two to evaluate, sometimes an afternoon, and sometimes it's a ten-minute-per-dog judgment that calls for more head than heart.

Paula describes watching the dogs' response to her as she walks between the cages. Who has bright eyes and perked ears? Who moves toward her with a confident, upbeat gait or a calm walk? Whose tail is wagging? Who stretches forward to sniff and, even through fencing, petitions for petting? The situation isn't always fair. In overrun, impoverished facilities, when stress on the intakes is high, sometimes even good dogs become reactive.

Paula feels the time and long experience in her knees when she kneels down on cement and extends her hands flat before kennel fencing, a hairsbreadth away from the chain link. She doesn't stare; she speaks softly, her voice low. *Show me who you are.*

Sometimes she'll go weeks before finding even one dog that seems right for service of any kind. Sometimes there are surprises, and a high-traffic city pound — a cinder-block noise box — will produce a handful of dogs that show their best selves on greeting, on leashes down the walkway, and out the door. Big dogs and small ones. White, black, brown, and every mottled variation in between. These are good pet candidates with something extra. They're highly motivated. They will sit on command or show they're up to learning it quickly. They'll bring a ball all the way back for a throw. They'll find a hidden treat. They'll submit to human handling and still come up cheerful. They are outgoing and aware. They don't snap at other dogs in passing. And, most important, they like humans a lot.

These are the possible service, detection, or therapy dogs, and in the space of just a few minutes, they've earned their way out of the pound. They will be scheduled for further testing and training, potential matching for jobs, and, if they do not measure up, re-homing as pets after obedience training has been completed. So there is life for them ahead. Good life. Activity, discipline, care, and kindness.

In the pound, Paula double-checks the numbers on the cages against the numbers on the collars. She notes their names, sometimes supplied by surrendering owners, often created by staff members who hope a great name might seal the deal on an adoption. Paula describes one rare day when she found a couple of mixes and a purebred: Dolly, Peanut, and Bud Lite. Dolly was a Chinese crested with a very good nose; Peanut was a nimble terrier with great dexterity; and Bud Lite — well, he might have been anything. A small, sociable retriever crossed with a cocker spaniel, he looked like he should be in a movie called *Honey, I Shrunk the Golden*. All of the dogs had crates, collars, tags, microchips, and foster homes waiting.

And then there was Jasper. Jasper was going home with her.

Jasper was an old hand at being rescued. He'd been adopted in another town three years before, a shelter success story. In 2007, puppy Jasper became the big-city lap dog of an elderly owner who loved him a great deal but whose illness moved faster than anyone had been prepared for. When his owner died, Jasper lived briefly at his vet's office, and for a time it appeared he might have a soft landing, the late owner having a son and daughter-in-law in the same city. Jasper moved in with them. But the daughter-in-law was pregnant, and Jasper's rich, curly, multicolored coat went everywhere. She was also very much allergic, and Jasper had to go. No one the couple knew wanted a dog, and they learned quickly that posting free-to-a-good-home listings on the Internet brought out only the creeps.

With reluctance, they took Jasper to their city pound, where polite but overworked volunteers told them they could make no promises, but they would try to find him a home. Small dogs had a little more hope, a higher adoption turnover; they got longer at the shelter. The couple left with these assurances and never inquired again. This time around, homeless Jasper was on day 43 of 45, a red tag on his cage, when Paula evaluated him. He looked like a good pet prospect for someone who liked offbeat charm. He was a friendly, spotty guy who didn't seem to find anyone a stranger, a terrier–Shih Tzu–poodle–something that defied even experienced dog fanciers' guesses, uncharitably called "fugly" by the people at the pound. Jasper looked like a jack-o'-lantern covered in dryer lint; he had a pronounced underbite and good teeth.

Sitting in the Port-A-Crate beside Paula on the drive home, Jasper exhibited the same quiet awareness of his surroundings that he had shown during the test. He sat upright in the crate, peering out the window. He was interested but not overstimulated, calm without being subdued. When she rolled to a stop and turned to look at him, Jasper looked back with the steadiest eye contact she had ever seen from a dog, his expression slightly expectant, as though he were waiting for some good word he had known in the past. Paula didn't know the word, and at the same time as she felt the bittersweetness of his condition, she was also aware that this steady watchfulness was a gift. Softened by the curly hair around his face, Jasper had a thoughtful gaze rather than a stare — a great quality.

The trip was a long one. Paula glanced in her rearview mirror across the expanse of travel crates. A couple of the dogs were restive. The other was asleep. The SUV smelled of kibble breath and wet dog fur. A few were encouraged enough to groom themselves a little: *Fff-pfft-ffff, flup-flup-flup.* One of the rubber hands was sticking up out of the bag against the window, looking like a kidnap victim's desperate attempt to get out. Paula says she always winces when she imagines explaining that to a state trooper.

As Jasper acclimated to the drive and the day's events, he watched Paula's movements in the car, and he tilted his head with his brow furrowed a little, as though he thought he could learn to drive, sure. As though he thought he could critique her driving, certainly. No matter what was really going on in his dog head, Jasper proved that he was curious, attentive, and not easily distracted.

Paula was smitten. She found herself thinking, like she always did, that if assistance or therapy partnership didn't work for Jasper, she'd take him. It is an old argument she has with herself every shelter-pull day. She has three dogs of her own, one very feeble, not one of them young, and Paula is on a fixed income, so it's not like she can afford more vet bills. But sometimes a dog moves her to the point of commitment. Jasper was that dog.

She switched on the radio to stay alert: classic rock and oldies, a syndicated DJ out of somewhere else. Jasper's ears perked. She could feel him watching her still, but there was a quivering edge-of the-dance-floor quality to his gaze now, like something in the music gave him pleasure, like *Baby, are you ready to rock?*

She says he had to warm up to it; he shifted position in his crate, then squeaked once or twice before he howled — a tenor croon that didn't rise to a squeal in the way of most dogs' howls. It was a practiced sound, pleasant and assured, an *ooo-OO-ooo, OO-oo-oo*. Paula laughed so hard her hands shook on the wheel, and at the precise moment she began to wonder if he would howl through the entire song (and through the entire trip, if she kept the radio on), Jasper took a breath through that curious underbite, deep and long, as though he was setting up for another righteous croon. The air whistled through the tight spaces of his teeth and made a single note.

Jasper could whistle, and Jasper could sing. *Ooo-OOO-oo, oo-WAWA-oo, sweeeeeeeeeeet.* The dog lover in Paula was uplifted. The dog evaluator's mind raced: *Just what might we do with a talent like that?*

Paula says she thinks about Jasper every time she gets discouraged. She tells others about him to demonstrate that great dogs can come in odd packages. She agrees to offer reality checks and be a sounding board for me on this business of finding homeless dogs with talent for psychiatric service, emotional support, or therapy. She's about to retire from all of it, though. Her sight is getting bad, and night driving has recently become much more treacherous. Paula says she might be a little hard to catch, but she recently got a smartphone. If she can figure out how to use it, she can send e-mail. She also got bifocals, so she can read it. The bifocals may be a good thing. Until now, Paula's poor vision led her to type in all caps, giving her message board posts and e-mails a Ten Commandments quality that set a few people on edge. This last message is no exception. Perhaps with Jasper in mind, she writes to me:

ANY DOG CAN SURPRISE YOU.
BEFORE YOU CAN FIND MANY DOGS FOR THIS WORK,
FIND ONE.

# 8

*THE SPIRIT OF SERVICE BEGINS AT HOME* — I remember a laminated poster of a child spooning soup into her toy bear's mouth in my fourth-grade Sunday-school class, a sort of Norman Rockwell–meets–Successories message probably lost on all of the students. Funny that I should remember that poster now, with a dustpan in one hand and a broom in the other and not a single living creature in the house inclined to help me clean. Pomeranians Mr. Sprits'l, Fo'c'sle Jack, Sam, Smokey, and Misty, plus Puzzle, the search dog: my dogs are not only *not* helping but actually making it harder. Housework involves a kind of hopscotch over and around them. The Poms are all semicomatose, strewn across the cool wood floors of our 115-year-old house, woolly and inert as bread mold. Though I can't see her, Puzzle is likely belly-up somewhere beneath a ceiling fan.

Dog lovers often talk about pets' untapped potential, and as I look down the corridor at this lot, I wonder where my own dogs fall on the scale of service gifts. With the exception of Puzzle, I suspect pretty low. They're nice dogs, but *honestly*.

"Practice evaluating dogs," some of my advisers recommend. "Practice on any dog that'll let you." It's good advice. Canine evaluators agree that dogs can miscue off a hesitant human. I look down the hall toward the Poms. I know! I can practice testing on *them*. How about a pop quiz for us all?

For working-dog candidates, a willingness to come when called is one of the most basic test items. In evaluating untrained dogs, we look for a dog's innate social attraction to humans. In evaluating trained dogs, which mine are (mostly, somewhat, maybe), the call and the friendly beckon examines obedience as much as attraction. I'm tired of cleaning, and maybe a little resentful that this crew is sleeping and I'm not, so I decide it's time for the dogs to show me a little action. Trained, untrained, whatever. "Hey, guys!" I call brightly and clap. "Hey! Hey!"

This would be the friendly beckon. I'm in the kitchen where the food is, so I should have some leverage.

A single Pomeranian ear flickers, but not one of them moves. Terrific. I am socially unattractive.

"Hey, guys! Pop quiz!" I try again. "*Come!*" This would be the obedience command.

They usually come. This time they don't. Granted, I'm a known quantity on a summer afternoon in the house we share. A stranger might have gotten more interest.

From another room, there's the heavy thump of paws on floor, and I see Puzzle rounding a corner and heading sleepily for me down the hall. She is the youngest of my dogs and, as a search-and-rescue canine, the most thoroughly trained. She's not hurrying, but she's not dawdling either; she winks in the bright light of the kitchen as she moves to me, her tail swaying lazily. When she gets to my feet, she sits and looks up patiently, as if unreasonable commands at odd hours are what she's used to. Her expression reads: *I was napping, but I'm here, and, you know, whatever it is, I'm in.*

I hear the awkward scramble of small feet, and behind Puzzle comes little Misty. The black-and-tan Pomeranian with white paws and crippled back legs is wobbling toward me as if swing dancing, her sweet face alight. It takes her some time to make it down the corridor, but she looks up cheerfully when she gets to me.

"Good girls!" I say to Puzzle and Misty, murmuring something about "Women rule." I rub Misty's ears and scratch her chest, hug Puz and kiss her muzzle. Misty sits with a panting grin; Puzzle melts into the hug, her tail thumping as she slides down to lie at my feet.

"You are the only good dogs in the house," I say loudly, pointedly. I move to the treat drawer. Puzzle lies where she has melted, watching me with liquid eyes. Misty pivots on her bottom, holding her Sit. Her little tan eyebrows go up. At the scrape and rattle of the treat drawer, the other Pomeranians suddenly rouse. They are *awake! awake!* and I am suddenly popular: The Caller of Dogs. The Giver of Good Things.

The Poms hurry down the hall with lively expressions of interest. They make a ring around my feet and sit, supremely good dogs all.

None of us are fooled. This is less social attraction or obedience and more *What's in it for me?*

Okay. So spontaneous testing at naptime is not entirely reasonable, or accurate, or fair. Top marks for Puz and Misty. I'll try the other Poms again at a better hour, and I remind myself that, all kidding aside, there's information in this experience: *when* I meet the coming homeless dogs may be just as important as *where* I meet them.

Evaluation of dogs for any reason is not an absolute science. There are always factors to consider. There are all kinds of evaluations for dogs at every stage. There are puppy tests that look at raw personality traits and drives; there are tests for adult dogs, tests for adult dogs in rescue, tests for dogs with special needs, and tests for grown dogs with known behavior issues. There are alternative test items for dogs with unusual backgrounds.

For potential working dogs, there are testing protocols that seek to identify dogs that might be more suited to protection or search than service. I've been studying every evaluation procedure I can find to learn what each contributes to understanding a dog, and I'm also looking for something more — possible test items that suggest not only willingness to partner a human but also an inherent connection, even concern. We can identify dogs that like other dogs better than they like humans. We can identify dogs that prefer humans to other dogs. But is there a way to recognize a dog's natural awareness of the human condition?

After the first test of attraction — the dog's willingness to come when a stranger beckons — the next item often evaluates the dog's interest in following a human who talks, claps, or attempts to engage him while in motion. Rolling a puppy over with a hand to the belly, establishing a pattern of strokes on an adult standing dog, or lifting a dog up and holding him aloft then examine the dog's willingness to be dominated by a human. Interest in working collaboratively is tested through a simple retrieval of a thrown ball or wad of paper. And finally, there is the question of stability. How does the dog respond to new stimuli? The dog is tested for touch sensitivity, sound sensitivity, and the reaction to an unexpected stimulus like a towel suddenly tossed in his direction. A good service candidate is alert and aware but doesn't hyperreact.

If I hope to help handlers train their own psych service dogs, I need to experience the partnership as closely as I can. First *find one,* Paula said. I'm looking for a dog to take through the whole process. I'd like to find one that can work as a co-teacher, help other handlers train their own dogs for service. Dogs definitely watch one another. Their human partners also need evidence that difficult tasks can, in fact, be taught. The demo dog can serve both functions.

Paula's recommendation is smart. Finding and training a dog will take me through the entire procedure every service dog handler faces — joys and failures alike — and if this work is something I'm just not up to, it'll be obvious here. This is a weed-out process for me as much as for the dogs.

I compare the formal evaluation procedures of Campbell, Volhard, and C-BARQ, and I talk to evaluators who use them or have created their own. Many of the test items are the procedures we use to test a would-be search K9, but for service partnership, we are looking for a dog that's different in important ways.

The following test items begin to sketch a portrait of a dog's nature. The scale doesn't determine whether a dog is a good dog or a bad dog. In the simplest terms, this scale suggests where a dog's impulses, affinities, and vulnerabilities lie. Highly independent and/or dominant dogs are often most suitable for search and detection work. Fearful dogs that lack confidence would not be appropriate for assistance or for search or detection. Search and service candidates typically show themselves as cooperative, engaged with humans, not overly aggressive but not passive or timid either. For assistance, we look for a dog that's enthusiastic, engaged, stable, and calm — a dog that shows balance, yes — but we also look for a dog that prefers human connection to any other thing. "A super-confident Velcro dog," as one trainer puts it. This kind of dog may well be a strong service candidate.

I look again at my own crew, and though I think I know how each would fare on the testing, I can't be sure until I try. Across days, I persuade neighbors, friends, and the occasional repairman to help me uncover any hidden talents the dogs might have. Search dog Puzzle's skill set is known, so I concentrate on the others. Each dog is tested singly, with the others behind a closed door. Because some of my dogs are homebound due to health issues, I don't take them out to neutral spaces. Testing them in their own environment isn't the cleanest slate possible — some of them may react protectively on their home turf — but this is all about hypothesis, and if nothing else, it will give me trial runs on test procedures. It will also tell me how well I know my dogs. Not one of them is a puppy. Most of them are rescues, and all of the rescues came with baggage.

The evaluation looks first at willingness. How eager is a dog to do this thing a human asks? Responses range from Unwilling to Reluctant to Compliant to Responsive to Enthusiastic. In many cases, those simple descriptors aren't enough — some dogs may be reluctant out of fear, others out of independence — so thoughtful observation and additional notes make a difference to our understanding of them.

## Testing Human Attraction

A repair contractor who likes dogs agrees to help. He comes into the kitchen after an afternoon's work in another room. The young man kneels in front of the dogs' beloved treat drawer but without a treat in hand; he claps and individually calls each of the dogs to come.

*Mr. Sprits'l, Pomeranian, age nine:* At the call of his name, from the end of the corridor, Mr. Sprits'l spins and barks and spins and barks and spins. And sits. And barks. Come? Not on your life. He'd score Unwilling. (Note: With a relative stranger on his own turf. On a genuine test, I'd run this one in a neutral space, but I suspect Mr. Sprits'l's score wouldn't change. He is a law unto himself, is Mr. Sprits'l.)

*Fo'c'sle Jack, Pomeranian, age ten:* Barks from one position, then comes forward, stops, barks, then comes forward all the way, wagging. He warms up, but it takes a bit. Because the final approach is cheerful (ears perked, wagging), he is Compliant.

*Sam, Pomeranian, rescue, age ten (?):* Sam is a partially blind rescue with a heart problem and a bullet in his back leg that is too surgically risky to remove. Despite hardship and abuse in his early life, sweet, friendly Sam comes readily forward at the call, snitzing (a sneezelike sniff) and smiling, ears perked and wagging. He's Enthusiastic.

*Smokey, Pomeranian-Chihuahua mix, rescue, age seven:* Sensitive and reactive, very acutely aware of eye contact from humans, Smokey is intelligent. But Smokey overworries, overthinks, overresponds. Still anxious after the loss of Erin, Smokey is wary of change, of strangers in general and of men in particular. The contractor is no exception. When called to come, Smokey stands his distant ground at the end of the corridor and barks in alarm, ducking out of sight and returning now and again to bark more. Smokey is Unwilling.

*Misty, Pomeranian, rescue, age ten:* Pretty Misty, overcoming her wayward, rolling gait, is eager to meet and greet. Her condition doesn't seem to dampen her friendly interest in other people. When the contractor kneels to call her, Misty scrambles down the hallway cheerfully, eyes bright and tail wiggling, then flops at his feet with an adoring gaze upward and a kiss for his fingertips. She is universally friendly and a consummate flirt. (*I am wonderful and you are amazing; together we are fabulous!*) Misty is Enthusiastic.

## Testing the Follow

Still testing each dog privately, the contractor gets up and walks away, calling the dog by name and gesturing him or her to come. Which dogs follow him? Which stay put? Which are completely alarmed by his movement?

*Mr. Sprits'l:* Barks from his unchanged position at the end of the hallway. *You moved!* he announces. Spins. Announces: *You moved! Augh! Augh! Augh!* Mr. Sprits'l remains Unwilling. When I put him behind the door, he's still fractious. He puts his nose to the crack at the bottom and snorts, *Hmph.*

*Fo'c'sle Jack:* When the contractor walks away, Jack follows from a distance. Sits at a distance. Watches. Looks at the contractor. Looks at the treat drawer. Holds his Sit, eyes bright. Snitzes. The closer the contractor gets to the treat drawer, the more Jack snitzes. Some behaviorists suggest the little sneezelike snitz means "I like what you're doing." Everything about Jack suggests he would follow this stranger, but for a price. This is not about the interest in humans. This is about the treat. There is no way I can credit Jack with compliance. The Pom with an ulterior motive, Jack is Unwilling. Jack's a sweet boy, but he's always got an angle.

*Sam:* Follows at a polite distance, his good eye turned to the contractor, tail wagging gently. They are perhaps three feet apart. Sam doesn't try to close the gap between them, but he's clearly interested in what's going on with the human. Sam's reserve is Responsive rather than Enthusiastic.

*Smokey:* Follows warily, at a distance, tail up, tail down, tail up, tail down, barking, barking, barking. Smokey's behavior straddles interest and caution, but he is more reactive than friendly, at least on his home turf. I'd call him Reluctant for now. It would be interesting to evaluate Smokey away from his home ground.

*Misty:* The Pom with the hardest time walking is also the Pom who follows most readily. Snitzing, smiling, and wagging, Misty follows the contractor's path across the kitchen, delighted with him. She play-bows, front paws low, backside high, and snitzes again. She is all about the charm and less about the treat. Misty is Enthusiastic.

## Testing Restraint

A friend who rarely visits the house attempts to see which Poms will allow her to gently hold them in place for one minute. Following recommendations from a number of canine tests, we will note who struggles and then submits with eye contact; who struggles without aggression; who struggles and attempts to bite; who totally submits and withdraws from human eye contact completely. This test is a basic indicator of a dog's willingness to submit to human control. Total submission is not the point here — assistance dogs need to have some self-reliance and independence about them — but a dog that will not give up the struggle, or, worse, struggles with an attempt to bite is likely not a good candidate for assistance work. None of my dogs bite or bare teeth. They have never snapped at strangers, groomers, or the vet. This is the kind of test item one has to do very carefully with an unfamiliar dog. My friend will gently but firmly restrain each of mine. I am sure there won't be bites, but I do expect some drama.

*Mr. Sprits'l:* Nothing doing. Sprits'l resists even the initial touch. Although at the vet, Sprits will submit to treatment with stoic tolerance, socially, in his own home, he is not about to be held in place for any reason, thank you very much. I suspect Sprits'l would never bite unless physically threatened, and maybe not even then, but he will also not acquiesce to this . . . this . . . *imprisonment* by a stranger. He struggles and mutters the entire time, giving us the stink eye. We try to hide our laughter, and we call his outrage Unwilling.

*Fo'c'sle Jack:* Struggles, submits, mutters in protest with eye gaze, then submits and tries to capitalize on the hold, leaning into the hands to get scratched. When in doubt, work it. Jack is Responsive, and if petted to his specifications, he's Enthusiastic, the little opportunist.

*Sam:* Submits to the restraint without eye contact. Though he doesn't struggle, I feel his unease. Sam's hard backstory gives me the sense he's been through a great deal. We do not prolong this. His behavior is the exact opposite of Sprits'l's. Sam's stoic but withdrawn response is Reluctant. I'm moved by his gentle acquiescence, but something in it saddens me for little Sam. (I'm sure he doesn't understand why he gets a secret treat later.)

*Smokey:* Struggles, resists, turns his head as if to snap without actually snapping, and, released, runs away. Smokey is an uncertain dog at home, sometimes fiercely independent, sometimes extremely shy. Curiously, he's a gentle, docile boy at the vet, submitting to thermometers and shots with a withdrawn quality much like Sam's. I believe Smokey's response here is less about dominance than about anxiety. He's Unwilling.

*Misty:* Struggles a little, as if startled by the restraint, settles, gives eye contact, and then seems to enjoy the hold, snitzing now and again. She leans into the hold until she melts, her belly upward to the cool. *You could scratch my belly.* Like Jack, Misty moves from Compliant to Responsive to Enthusiastic, and we wonder aloud just who's testing whom in this exercise.

## Testing Dominance

Dominance is often evaluated by testing how puppies react to being held off the ground by a human. With adult dogs that are too large to pick up and hold aloft, evaluators sometimes test dominance by extended firm petting, by lifting the dog's lip with a finger, or by the hand-over-paw test: covering the dog's paw with a hand, holding some pressure, and seeing how the dog responds. This kind of testing is not done carelessly with unfamiliar dogs. Because the Poms are all used to getting their teeth checked, the lifted lip, even from a stranger, isn't the fresh experience it might be for other dogs. We try the hand-over-paw.

*Mr. Sprits'l:* Insulted, withdraws his paw and dashes off, then barks indignantly from ten feet away. Sprits'l is Unwilling! Seriously Unwilling! Again!

*Fo'c'sle Jack:* Wiggles, licks hand, withdraws his paw, play-bows, and extends his paw again. Jack, like many Poms, likes the game grab-toes, and even from a stranger, this test is an irresistible call to play. Jack is Responsive, even Enthusiastic. When his paw is held down with firmer pressure so that he can't withdraw it, he's Compliant — but he snorts and tosses his head a little. He looks vaguely disappointed that we don't follow the rules of the game.

*Sam:* Allows the hold of his paw, lies panting slightly nervously, his eyes far away. Sam has withdrawn from us without being able to leave. He submits, but there's old, old uncertainty in his expression. We quickly release him. Sam is Reluctant.

*Smokey:* Alarmed by the capture of his paw, Smokey struggles, withdraws the paw, and runs away, barking from under a bed. Smokey is Unwilling.

*Misty:* Wiggles, licks the hand, withdraws the paw, snitzes, and puts her paw on the hand for another round. Like Jack, Misty loves grabtoes. She is all joy at the attention, and she is Enthusiastic. When a firmer grip is held on her paw, she allows it, bending to give the evaluator a single lick.

## Testing the Retrieve

How interested is the dog in working with a human? Though there are other ways to assess this with grown dogs, a quick and common test is to toss a crumpled wad of paper or a ball and see what the dog does. Does the dog chase, pick up the thing, and run away? Chase, stand over it, and guard? Chase, bring it back? Chase, let it go without bringing it back? Lose interest quickly — or not show any interest at all?

I'm interested in the results of this with my crew. Poms are not typically considered retrievers, but they are a trainable, playful breed. There is herding and performance in their history. How well does a ball toss test dogs that prefer other games? We may need to develop measures of canine collaboration that don't involve retrieving. I toss a ball, and if that gets no response, I toss a wad of paper.

*Mr. Sprits'l:* Barks at the ball, warily approaches the wad of paper, sniffs it with a suspicious eye in my direction, and trots off. Sprits'l is also Unwilling.

*Fo'c'sle Jack:* Ignores the ball, goes after the wad of paper, sniffs to see if it's food, looks balefully at me when it is not, and stomps away. On the common standard for this test, Jack is ultimately Unwilling. He is not interested in retrieving.

*Sam:* Scurries after the ball on his aged, arthritic legs, captures it with his paws but does not attempt to retrieve it. Sam looks back at me, panting, then flops down beside the ball in a bemused rather than guarding posture. I think his energy has failed him. There's nothing possessive about his posture. After a moment, he gets up and returns to me, a little crestfallen, as if somehow the game failed or he failed the game. Sam is Responsive, nonetheless.

*Smokey:* Springs barking after the ball, picks it up, and returns it, but more than that — he glows. Smokey drops it easily at my feet and retrieves as many times as I throw the ball, tail up and happier each time than the one before. He spins and circles after each retrieve, thrilled. Smokey is joyfully Enthusiastic.

*Misty:* Yaps happily at the ball but does not attempt to chase it on those wobbly back legs. The ball simply moves too fast. She scampers for the tossed wad of paper and returns it, dropping it at my feet, grinning up for a second toss, which she also returns. Who knew Misty would also fetch? High gamesmanship from a senior, disabled Pomeranian. Misty is also Enthusiastic.

## Testing Stability

How do the dogs react to an unexpected stimulus, such as a dropped object or a sharp noise? From working dogs — even from pet dogs trying to pass the Canine Good Citizen test — we look for awareness and intelligent interest without overreaction; wariness, perhaps, but without marked fear. Here the scale is also about the degree of response, from a dog that rushes and barks at the stimulus to a dog that runs away and hides. A nonreactive dog usually listens and locates the object, perhaps also moving toward it to investigate. For our needs, we like a dog that obviously notes the stimulus and responds, recovering quickly, if startled, or inspecting curiously.

When the dog isn't looking, we drop a large book nearby. This scale ranges from Afraid to Provoked to Wary to Curious to Responsive to Uninterested.

*Mr. Sprits'l:* Jumps, turns and barks happily at the book, then comes forward to sniff it. Mr. Sprits'l, though opinionated and reactive in many ways, gets a solid stability mark for his caution combined with curiosity. He is Wary and then Curious.

*Fo'c'sle Jack:* Turns and briefly looks at the object. While Jack often scurries around big dogs or people who might step on him, a dropped book is an object of only mild concern. Jack isn't sound sensitive to this kind of stimulus. He is Responsive.

*Sam:* Turns and looks at the object without moving. Looks at me mildly, like, *Well,* that *was different.* Sam is Responsive.

*Smokey:* Skitters, turns, and barks in alarm. He certainly backs off, but he doesn't mind telling off the book a little. Reactive Smokey is Provoked.

*Misty:* Twitches a bit at the sound, approaches and looks at the object, then looks at me, panting and grinning. Misty is Curious.

Thinking about seizures, panic attacks, hyperventilation, and the like, I am interested to see how the dogs respond to an unusual human stimulus. Some dogs are easily frightened by human behaviors they don't understand. A friend the dogs know fairly well agrees to be a volunteer victim. In our backyard, without warning, she falls to the ground in the presence of each dog, one at a time.

*Mr. Sprits'l:* Barks, runs over to the victim, inspects, barks again. Sprits'l is Provoked by the fall, but he's also Curious. I'm interested in the fact that he remains beside her, spinning and barking, rather than retreating and barking from farther away.

*Fo'c'sle Jack:* Jack gives me the response I had expected from Mr. Sprits'l. He stays where he stands, barks a few times from that position, then turns away. *Human on the ground? Humans are strange. So be it.* Jack's distant bark is Provoked, but he is a verbal Pomeranian, inclined to comment at least once about many things. Jack is mostly Responsive and then Uninterested. Would a treat in my friend's hand have made a difference? Oh, you bet.

*Sam:* Toddles carefully over to the fallen friend, sniffs her thoughtfully, then sits beside her. He is Curious. He is also the first dog to choose to sit close to the human on the ground.

*Smokey:* Barks in alarm, approaches halfway, barks more, approaches cautiously to about three feet. Comes no closer. Smokey is Provoked and Afraid.

*Misty:* Head perks, tilts. She scampers over to sniff, then flops down beside the person. She is not alarmed, but, like Sam, she seems to feel it a duty to stay nearby. Misty is also Curious.

The trial evaluations have been interesting. Using this casual hybrid of tests, I find Misty alone shines in the areas suitable for service. Some of the other Poms excel in this category or that one, but it is Misty who possesses the interest, engagement, and stability we look for in a service candidate. *Any dog can surprise you,* Paula wrote. I've always thought Misty was a sweetheart, overshadowed in her former home by her adopted sibling Smokey, but even I would not have predicted how well she would do. Misty is a good example of a dog with right-on aptitudes and a troubled body. She is a motivated little lap dog, age ten, with severely impaired back legs and breathing issues. Though I am looking for a demonstration-dog candidate, Misty's age, disabilities, and current health make even demonstration service work unrealistic. But her connection to humans is strong. Even senior animals can work as therapy dogs in hospitals and schools, and with her outgoing nature, Misty would love this. I see Misty's joy in human contact with new eyes, reminded that too often we let a frail dog's disability define her, overlooking all that is offered by an eager, affectionate heart.

# 9

Sympatico — affinity — a bond, the handler tells me. In his opinion, Gene says, great aptitude tests won't add up to a good psych dog partnership if there isn't a strong connection between the human and the dog. Gene is a handler whose canine partner is about to retire, a handler whose condition was so markedly improved beside his service dog that when we speak, he does not think he will need to find another.

A lot of people talk about "the bond." Not everyone agrees about its meaning. "Is the bond about *love?*" I ask, needing his definition. "In your opinion, do you have to love the dog first?"

Gene says no. Affinity happens when you like who the dog is in his bones and something about you in your bones the dog likes too. Affinity happens when you'd choose to be together voluntarily. It can happen quickly or take a little longer. He's not sure what makes up this attraction, but he is sure it's one of the reasons why you see so many different kinds of dogs in psychiatric service. Big, small, energetic, sedate, earnest, soulful, or slightly clownish. Look at the pictures on the web forums, he tells me. There's a whole dog world of service out there. Not everyone connects to the same dog.

Gene firmly believes that the psych dog handler should have a voice in the choosing of the partner. He puzzles a little over the right way to say it, but the gist of it is that mental conditions evolve and symptoms change, and you and your dog have to adapt to that. The dog ages and changes too. Gene wonders — how can the two of you adapt to changes if you don't really like your dog or if your dog doesn't really like you?

I bring up the "arranged marriage" that began my working partnership with search dog Puzzle and the strong connection that developed after a very rocky start. She was chosen for me by her breeder and my search team's head trainer, the choice based on two sets of test scores and the breeder's observations of Puzzle from her birth. Affinity, love, trust — all those good things are there now, but they certainly weren't at the beginning. Would I have chosen her if I'd met the whole litter? Probably, based on her test results, even though she didn't seem to much like me. Her aptitude scores were high. If she'd been given a choice, I have no doubt that in a lineup of handlers, I would have been last on Puzzle's list. But it did work out, somehow. There are plenty of service dog organizations that place trained dogs with partners the animals have never met, and the famous bond builds (or, in occasional cases, doesn't) as each pair begins to train for their lives together. Affinity grows over time.

With psychiatric assistance dogs, does a dog have to like you (in dog terms) to help you? I wonder. Do you have to like a dog to be helped? What kind of bond must exist in that particular service relationship? There are strong voices on both sides of these debates.

Gene and I talk about the news stories where some random dog made a heroic move for a stranger — alerting to danger, protecting from harm, warming the fallen in the middle of a snowstorm — a one-time rescue of brilliance and compassion. We agree that dogs have the ability to surprise us, but Gene goes on to insist that the daily saving works only when a psych dog and his partner have a relationship built on something more than tasks and treats.

I ask him how long it took to choose his dog. Gene smiles, shakes his head, and says he was all kinds of lucky. His dog chose him.

Gene had always been a worrier. That's what his mother said. From the time he was little, he was sensitive, got knotted up by things that didn't bother the other kids. When his brothers were fascinated by the death throes of a bird that flew into the glass of their patio door, four-year-old Gene cried, imagining the pain, however brief, of the bird's broken neck. Gene was a quiet kid with an internal life his family didn't quite understand. He read early and willfully, staying up long past bedtime to read in the glow of a streetlamp outside the window. He was spanked for that several times. He screwed up his eyesight.

Gene remembers himself struggling through school mostly untouched by education, unconnected with his peers. He can recall a pretty young student teacher who took a special interest when he was a high-school freshman. She had read *Cipher in the Snow*, she mentioned one day. Had he read it? she wondered. He hadn't, but he did, and afterward he got the sense that her encouragement wasn't due to any real interest she had in who he was but rather to her determination not to be the teacher in the story, who had nothing to say at a boy outcast's funeral. In a teenage culture where despising was very much the thing, Gene let her off the hook. He didn't despise her for this. He didn't despise anyone for anything. He mostly just read, standing with a book in the crossroads of his mother's menopause and his father's business troubles, his brothers' unfettered holiday through girls, jobs, and alcohol.

Gene remembers his weeping mother bowed over his oldest brother, whose stomach had had to be pumped after a night with friends at the lake. He'd been hauled into the hospital comatose. When he woke a day later, sunken-eyed and vague, the best his brother had in him to say was only "Shit-I'm-sorry-Mom." Then he high-fived the next-eldest when his mother turned away. "Fuckin' messed up," the two had whispered to each other, grinning. Gene didn't despise his brothers, or his mother for being blind to their faults, or his father for always being at the store. He just worried that those brothers would fucking mess up enough that they wouldn't be able to take over the store the way they were expected to — they were the social guys, the front-of-store guys — and that the family business would then fold, which was possible, and that his parents would die in poverty because Gene — definitely not a front-of-store guy — wasn't enough to hold it all together.

He kept quiet.

Gene's marriage, at twenty-two, had surprised all of them. *How the . . . how . . . Jesus H. . . .* , his brothers had seemed to say to each other. *How could a guy like Gene get a girl like Shannon,* a woman whose bookishness by no means got in the way of a body they would stop a train for, gladly. Gene himself was never sure how it happened, how a shared addiction to *Lord of the Rings* translated to an engagement, a marriage, and a baby (in that order, but it was a close call; this is what Arwen and Aragorn could do to you). Love, he was sure of. Shannon, he was not. Her remoteness in many ways matched his own.

She was a magical thinker. He was her project, someone told him later.

Was he shocked when Shannon left? Was he surprised she had stayed so long? What happened (or didn't) between them? He would never know. Shannon left gently, with a kiss before work and a letter, weary and absolute, on the table afterward. He knew her well enough not to argue. No part of her note was romantic; nothing in the hard angles of her printing said *Come after me. Fight for me. Prove me wrong.* They had settled things with the detachment of the drugged or the drowned.

Shannon left cleanly. Left not just him but their son too. At thirty-three, Gene found himself a single father to a son who was unlike either of his parents. He was open. Grounded. A rock-solid kid. He was Beren on the birth certificate, because they'd been pretty damn fanciful in those days, and Ben in practice, because later they thought they ought to give the little guy a break.

Gene was thirty-eight when he had his heart attack. No obvious cause: the stress of his recent days no different than the stress of any other days of his life he could remember. No warning; no numbness, no chest pain, nothing really until that moment on the highway, taking Ben to his busboy job, when suddenly Gene's heart seized and his veins caught fire. "Oh," he said, like a sob, then he doubled over reflexively, head to the steering wheel. He vaguely remembers the hard smack of it against his cheek. What he doesn't remember is veering into the next lane, the car that hit him, or the car he hit. He doesn't remember the rending of metal or the shattering of glass, the smell of hot rubber as the tires of all three cars slid wildly across the asphalt. He doesn't remember the screams of the injured, including his boy, his rock-solid boy shattered in a baker's dozen places.

Traumatic amnesia could have been a kind of grace, but in the absence of memory, Gene the worrier imagined all of it. Over and over. He held himself responsible. He refused to be absolved. "I could have killed someone," he insisted to his boy more than a year after Ben's last surgeries. "I could have killed you."

Gene realizes now that this too could have ended them. Not the accident, but the aftermath, when frail, disabled Ben was determined to push forward, and Gene was stuck in the hamster wheel of his own despair. Bedridden at first, then wheelchair-bound, Ben was determined to get on his feet sooner rather than later. Gene was at his best when caring for his son, but in all other ways, he withdrew. He could never quite get over the sensation that every heartbeat would be his last. He gave up driving. He couldn't focus enough to work, gripping the phone, the desk, the elevator doors, expecting his heart to fail. After a whole series of surgeons and doctors and psychiatrists treated him, Gene was declared disabled.

Neither of them would have imagined this was a good time to bring in a pet, but two doctors and an occupational therapist recommended a service dog for Ben, and Merlin's entry would turn both their lives around.

Merlin had known his own hard knocks. One of eight puppies, he was the sole survivor when he and his littermates were put in a garbage bag and dumped over the side of a low-water bridge in Connecticut. A rat terrier, straining on the end of her leash during a morning walk, found the bag of puppies. The terrier's owner had nearly pulled the dog away, suspecting old food or every kind of grossness, but when her little dog yipped and skipped, she gritted her teeth and opened the bag. And there in the middle was Merlin, an inkblot of puppy with a broken tail who'd been protected from the fall by his brothers and sisters, and there was the good woman with her insistent terrier who couldn't keep the pup due to housing restrictions. She tucked him into her windbreaker and snuck him home via the alley and a back door.

Merlin's escape was so miraculous that, for a little while, he was bottle-raised in secret, passed from neighbor to neighbor, always one jump ahead of the more self-righteous members of the homeowners' association. He should have been named Hush, they said it to him so much; he would probably have answered to that. He got bigger and louder and became a much harder secret to keep. Everyone asked everywhere, trying to find a home for him, and things were looking grim indeed until a breeder who raised puppies for service offered to add him to her litter of golden retriever pups that were about the same age. With any luck, the golden mama, a kindly soul, would take him as one of her own.

When the time came for the golden litter to go off to puppy raisers, Merlin's future again was uncertain. Nothing was known of his background. He'd had the benefit of early rescue, good nutrition, socialization, and vet care, but there was a whole host of unanswered questions. As just one more black dog needing a home, Merlin didn't have a whole lot of options. With eyebrows raised and fingers crossed, the breeder sent Merlin off to a service puppy raiser to see just what kind of dog he would be.

*Calm and friendly,* an evaluator wrote about him at four months. *No resource guarding. Soft-mouthed, but generous. Curious. Open to commands. A little passive. Slow: not as quick to learn as his peers.* Every word mattered. The lovability of the dog is not the issue, and a service evaluator's job is straightforward. Does the dog show promise or not? Her notes shaped Merlin's future, and those last two items, passive and slow, weren't winners. Unlike the program puppies who came from breeders, Merlin had no safety net if he washed out. What was it that caused her to evaluate him a second time — some deep likability or the faint spark of something bright in a dog otherwise considered a little dim?

It was Merlin's mouth, his obvious pleasure at picking up objects and his willingness to surrender what he picked up, that carried the day. For a time he lived and worked with an occupational therapist, spending his hours learning new commands and interacting with her clients, helping them regain their own dexterity. The dog was something of a goof, a class clown. His once-broken tail creaked when he wagged it, to the amusement of all. Merlin could have lived out his days in this good life and service, but then the therapist met Ben, and Ben's father, Gene, and worked Merlin with them both. For whatever reason, the dog lit up in their presence. He was quicker, smarter, more funny. And working with Merlin, the kid grew a little more confident. The sad-sack father laughed. After a consultation with Ben's doctors and much soul-searching on the part of the therapist, Merlin went to a new home.

The first thing they had to do was imagine life without thumbs. Merlin didn't have any, which forced Ben and Gene to reframe the household. Doorknobs were replaced with handles; the refrigerator was fitted with a tug rope. Slick things like the phone got temporary bands of masking tape as Merlin learned how to carry them. It was interesting, says Gene, to think like a dog, to imagine having only blunt paws and a mouth. Merlin followed Gene from room to room, companionably inspecting changes, adapting to his expanded duties. If the dog was dim, Gene says, it was probably a good thing. They were all feeling their way at this. The daily lessons gave them all something useful to do.

As for Merlin, he was much like his boy in temperament. He capered for approval. He really seemed to enjoy the service tasks Ben required of him. The dog padded slowly beside Ben's wheelchair. He slept by the chair or by Ben's bed, ready to pick up or retrieve anything Ben seemed to need. He was calm in the presence of Ben's pain, which could be severe, and Ben's depression, which often settled over him in the night after his pain had been most terrible.

Gene remembers the brilliant day Ben moved from wheelchair to walker. It was a physical therapy day like any other. None of them had expected it. Though Ben's arm strength had not yet been proven, all of a sudden he looked up at his father (not the therapist, his *father*) and said, "I'd like to try. Will you help me try?" And Gene and the therapist scrambled for a walker, and his boy was sitting in the wheelchair, biting his lip, and then he was forward, and then he was up.

When he stood up, Merlin, who'd been standing beside the chair, sat down.

"Good boy!" Gene shouted. Ben flushed. Merlin panted a grin.

You think you know what a miracle is the moment you see your son born, Gene says, you think you know the fullest measure of love, but then you see your down boy get up and make a shuffle that is almost a step, and born — as good as born is — is just a rehearsal.

It was the kind of triumph made for Hollywood. If only life moved at the speed of a movie montage, an ellipses of months of steady progress. Cue the music: Show Ben standing, walking, pushing open a door. Making a sandwich! Taking a shower on his own! More music, and the boy goes back to school, gets a job, kisses a girl. He no longer stares up from the bed with his teeth clenched, wondering if what *is* will ever change to something better. Gene wished that kind of upbeat passage for his boy. In truth, Ben's healing was a harder journey. For every day or two of gained ability, he had a day of almost equal losses. Patient Merlin continued to serve beside him, the "dim" dog constantly at the periphery, head up and creaky tail waving gently, always on call, some days needed and some days not.

When was it that Merlin's job description changed? Gene doesn't really remember. Ben noticed it first. As Ben improved and became more self-sufficient, as he regained balance and the ability to carry things on his own, Merlin moved from his sleep spot beside Ben's bed to a new position in the doorway. At first, Ben thought the dog was choosing to watch over Ben's new mobility. From the door, Merlin could follow his progress from bedroom to bathroom and back again. He could also look down the hallway with an eye to Gene.

Now Ben wonders if Merlin sensed a shift in need. Ben had grown stronger. Apart from his diligent caregiving, Gene was much the same. He still had panic attacks. He hadn't driven a car since the accident. As Ben's steps grew more sure, Gene's continued as though he were treading on glass. He shook sometimes like an old, old man.

Then there was the day they had an argument, Ben and Gene, about Ben following his doctors' orders and Gene pretty much avoiding the ones his own doctors gave, and how could Gene expect his heart to get stronger without trying, without exercise? "What will you do when I'm gone?" Ben shouted, and Gene felt the ground shift. He was thrilled his boy had hope enough to talk of leaving. He was scared shitless about what would happen if Ben left.

Gene pushed through the front door to prove to his son that, by God, he did exercise, and he stood there at the edge of the known world without knowing where to go or how to get there. He felt his heart clench and unclench like a fist. He glanced down at the bottom of the stoop and Merlin was beside him, looking up.

Generalized anxiety disorder. When doctors gave a name to the weight of worry he had carried as long as he could remember, Gene was somehow relieved. A thing with a name was concrete. It could be dissected and understood. (*Generalized* was a little problematic, though. Didn't that mean that anything could become a source of anxiety? Anything? How do you control anything? How do you control everything? When Gene realized he was worrying about the name of his condition, he remembers laughing. He says that may have saved him.) He had no clue how to walk out of the maze he'd been lost in most of his life. There was always this sense of doom, this condition of eclipse. He could not be sure if the anxiety caused the heart attack or was a harbinger of it. Meds he had and meds he took, and they probably made a difference. But in solid, undemanding, un-fanciful Merlin, Gene began to make his way free.

On-duty, Merlin was used to moving at the speed of Ben's wheelchair. Neighborhood walks with Gene were at much the same pace. One block and back again. Two. Ben's physical therapist helped Gene teach Merlin to lean into Gene's knees when he went shaky, a Brace command, precise as a dance step between them. Step, step, wobble, "Merlin, Brace" ("Good boy," ear scratch). Mix it up when you practice, a dog trainer told Gene. Make sure you aren't teaching him that after every two steps you wobble. And so man and dog practiced at home, with Ben in the recliner playing video games, peering up from the controller and offering advice between rounds of Gears of War.

Gene and Merlin began to walk twice a day. Aware that his own rigid thinking could slow down his progress, Gene tried to vary the route and the length of the walk each time. Merlin knew what *Brace* meant, and he came to recognize unsteadiness. Gene didn't always have to speak the command. The dog learned to anticipate it by feeling Gene's first wobbling step after half a dozen sure ones. He sensed the change of gait, the new tension on the lead, and Gene's hesitation. Merlin would lean against Gene's trembling leg, holding there until Gene's "Okay" released them both. Not such a dim dog, Gene says about Merlin. After every successful Brace, Gene would cup his hand over the dog's forehead, rub his thumb affectionately along his muzzle.

It was the not-dim dog who began to intercede at the onset of Gene's panic episodes. They began after the heart attack, after the car accident, and to Gene, they seem to come with the surprise of an assault — and they come even on good days, when things are going well. A tightness in his chest, his heart clenches and skips, he can't breathe. Gene says he reaches out and even solid walls feel fuzzy and insubstantial. Though Gene tried, unsuccessfully, to hide them from his son, the attacks seemed like they were never going away. They were still frequent and unpredictable enough that they prevented him from keeping a job, profound enough that they could drop him to the floor. Merlin found him there once, Gene crouching on his knees and hands in a desperate attempt to keep the bathroom linoleum from rocking. Curious but unafraid, the dog washed Gene's face with his tongue, then settled beside him for the duration. Gene rode the attack out with his nose to the linoleum and his arm around the dog. Merlin did not move.

Gene isn't sure what Merlin noticed — the hyperventilation; the shaky balance; a change of scent, perhaps, from his rush of adrenaline and cold sweat — but it didn't take long before Merlin was there at the first sign of an attack, standing by expectantly as they progressed, washing Gene's face if he fell to his knees. The dog began to arrive earlier and earlier in the attack, sometimes just at the moment Gene felt the first sink of spirits that meant it was all starting again.

Once, Ben, coming around the corner on crutches, caught them. Gene was gasping on the couch, hands digging into the upholstery, bent almost double toward the dog, whose tail gently waved, whose nose snuffled inquiringly across his face.

"Dad?" Ben said. Ben knew about his father's attacks. Gene thinks he knew about most of the episodes Gene tried to hide. "Dad, do I call 911?"

Gene shook his head, which made the room spin more — suede cloth and black-and-gold wallpaper wheeling — and just about the time he thought he'd fall off the couch, he wrapped his arms around the dog.

"Merlin," he whispered, and the dog held very still as he had done before, even though this was outside his formal training and apart from any command he had ever learned. Gene rested his cheek against Merlin's spine and closed his eyes, feeling the steady thrum of the dog's pulse and his slow, even respiration. Curious as all this probably was to the dog, Merlin wasn't alarmed. Gene held on, inhaling when Merlin inhaled, exhaling as the dog did. He remembers that first effort to stay upright by absorbing the dog's calm certainty.

Gene did not fall to the floor. He doesn't know exactly how long he held there, how long his son, standing on his good leg and his bad one and his crutches, watched his father pitch that private battle in the company of the dog. But the episode marked a turning point. Merlin could brace Gene's unsteadiness and loosen the hold of Gene's panic attacks, and as Ben needed him less, Gene owned up to needing him more. That day marked a wordless agreement, and Merlin became Gene's dog.

Not that simple, Gene's therapist said to him. He would not get off that easily. She agreed that Merlin was as useful for Gene as he once was for Ben. She was also in the business of urging Gene to be proactive in his own healing. The therapist had a dog herself, but she was new to the idea of psych service dogs. Nonetheless, she'd read up, and she was firm with Gene. At its best, this partnership is a process. The guy you are now is not the guy you'll be tomorrow. The dog you have now will change too.

She was ahead of all Gene's arguments. No matter that Merlin was already an old hand at this. He was highly obedient and a fine assistant in public. No matter that he already knew how to brace Gene standing and stabilize him during panic attacks. Gene owed it to the dog and to himself to train for the ongoing alliance. She wanted them to take some kind of class. The therapist wasn't particular. Merlin could learn to jump through hoops or walk on his back legs or bark "Jingle Bells" for all she cared, but she believed that better things would come if they had a common goal and learned more about each other to achieve it. She gave him that push, and she didn't let it drop the next week when Gene tried to sidestep the issue.

Gene wasn't fooled. She was urging him out in public with his dog, and he was going to have to walk, or ride a bus, or drive to get there.

Gene looked up classes on the Internet and was amazed — amazed! — at how many things he could do with his dog. Obedience classes he expected (and he was briefly tempted to go into the beginner classes with Merlin vestless, a ringer dog who aced all the commands and made them both look brilliant). There were trick classes using clickers like castanets, and at the little tick-tock sound, dogs could learn old tricks like roll over or new ones like high-five. There were nose-work classes, tracking classes, and agility classes. There were urban-dog classes where Gene could teach Merlin to skateboard.

Gene learned that in some cities, it got a little more fantastic. There were etiquette classes for dog tea paw-ties and sewing classes for dog costumes and ceremony classes for dog weddings. Dog *weddings?* Owners and dogs could do "doga" — or yoga — together. There were classes with animal communicators where they could learn to communicate psychically. (One night, for fun, he and Ben sat in the living room and thought hard of bacon, wondering if Merlin would give either one of them a glance. "Okaaaay. Major *fail*," said Ben. Merlin's tail thumped at the sound of his voice.) There was a whole dog culture out there Gene had never imagined existed. He could not find their place in it.

"Look to the dog," a friend in his counseling group advised. What would give Merlin the greatest pleasure?

It was a good question. Gene and Ben had long been determined that Merlin should have a dog life of his own. This hadn't always been easy. Merlin came to them already dutiful, easygoing, and amiable. Born to partner, he was a whither-thou-goest dog from the start, maybe a little less hey-hey-*heya* than one would expect from a Lab. He liked a ball well enough, but after a few retrieves he was done with it. Gene and Ben tried other toys Gene's mother found at garage sales, pre-chewed stuffed ducks and squirrels that Merlin shook a little and discarded. The first time Gene threw the dog a Frisbee, Merlin raised his eyebrows, and the disc hit him in the head.

"Dude," said Ben, doubtfully.

"Not his game," Gene replied.

But what was?

It took an elderly neighbor with a tan poodle in a tutu for them to discover it. The little vacant English lady with her littler dog was a visitor to the house next door, and the first time she let Pookie out, Merlin was also having an outing in his own backyard. New to the neighborhood she might have been, but Pookie took umbrage and rushed the fence straightaway in great oingy-poingy bounds, savaging the big dog from yards away. *My turf!* (boing) *My turf!* (boing) *Mine!* (boing) *Mine!* (boing) *Mine!* She was a tiny thing but outraged, a dedicated fence fighter, rushing up and down the chain link in her pink tulle and spangles like stars, yapping furiously at the black Lab twenty times her weight.

From the back porch, the old lady scolded Pookie in a wavering voice. From his back deck, Gene watched. Merlin could eat that little dog whole, without chewing. Gene had seen other dogs escalate hostilities during a fence fight. But Merlin seemed unprovoked. At first puzzled, then in some way amused, Merlin seemed to enjoy Pookie, tilting his head back and forth as if he weren't quite sure what to make of her, then wagging cheerfully and ambling along the fence line in sync with the poodle, huffing into the fence and revving her up again if she slowed, his easy lope to her full gallop. He didn't growl, and he didn't bark; his hackles didn't raise. *Creak, creak, snap,* wagged his broken tail. Pookie was some kinda something. She was relentless. He was charmed.

How long could they keep this up? Gene and Ben watched one afternoon as Merlin paced the fence line with the poodle until they were both tired. He flopped down before the fence, nonchalantly chewing a paw, while Pookie grumbled, nose through the links, pawing at the ground with silver toenails. She'd lost some of her venom; she had barked so much she was hoarse. In time, she extended her dime-sized nose to exchange a gentle sniff with him.

When the young woman of the house called the poodle inside, Pookie prissed away, curly poodle backside blooming out from under the tutu. Merlin watched her go with a besotted expression. The little dog disappeared, the door banged shut, and the big dog's face fell. He put his head on his paws and sighed. On later outings, he would gravitate to the same spot again. He was loyal to their meeting place, waiting for the poodle to come out and smack-talk him some more.

Merlin liked playing with little dogs.

A cautious trip to the dog park confirmed it. The first time they took him, Merlin ignored the big-dog arena with its fellow retrievers and shepherds and spotty mixes. He keened instead toward the fenced-off playground for the littlest dogs. Though there was a size limit for the small-dog compound, Merlin's service vest made a good enough impression that he won entrance to the world of the Lilliputians. Gene pulled off the vest, and Merlin was in heaven. The little-dog compound was full of puppies and grown dogs that came no higher than his elbows. He wagged and play-bowed, petitioning to be chased.

"They say horses don't know what size they are," said Ben, just throwing it out there.

"What a good dog you have," said an elderly man with an anxious, wild-eyed dachshund that happily shadowed Merlin as he circuited.

"You should have named him Paxil," added another. (Oh, the things Gene could have said but did not.)

Word got around that Merlin was good with small, reactive dogs. Somehow a casual volunteerism emerged. Gene and his service dog became regulars. They made dog-park dates at odd hours with others. Gene, the former recluse, now sits on a park bench with like-minded friends, cradling a cup of coffee, watching Merlin trot a circular course through the compound, a black sun to an orbit of little dogs of all tempers. The Lab can keep this up for hours, giving himself plenty of exercise and the small dogs four times more.

# 10

IT IS DARK, DARK, and I am dreaming of dogs when one of my own wakes me. It is Sam, I think at first, because I've been dreaming about him — Sam, who is known for his sweet shyness, his soft requests for attention. It takes me a moment to realize it can't be Sam, whose heart failed as he lay in my arms during a seizure ten days before. I resuscitated him only to have him die again a few minutes later. I wake to feel his absence like a bruise.

No, it's not Sam. It's Misty now at my shoulder pawing frantically. Was I having a nightmare? I roll over and recognize that we must have had a hard night. I am lying diagonally across the bed, with my head almost at the foot of it, and Misty is on my pillow. But something isn't right. She's lying uneasily on her chest, like a boat tilted to starboard, and she's got her neck stretched oddly. I can hear her gasping for air. In the faint light from a hallway lamp, her dark eyes are terrified.

How long has she been in distress? How hard was it to wake me? I don't know, but I can see the situation is bad, and worse, I can hear the wet, gurgly sound of her breathing. Misty is drowning. This is congestive heart failure at its ugliest. I've seen it before. I am up out of bed immediately, and with a glance at the clock I realize that the lowered window shades made it seem earlier and darker than it is. It is actually 6:45 A.M.

Shoes. Glasses. Driver's license. Keys. And behind the search for all of that, the gurgle of a little dog's lungs.

Misty.

And which way to travel? The emergency vet will close in forty-five minutes. A regular vet clinic miles in the other direction will open in forty-five, but vets may not arrive for another hour and a half.

From her position on the bed, Misty wheezes and watches me throw on tennis shoes with my pajamas, phone cradled against my shoulder as I call the emergency vet asking if I can give her more Lasix before we make our way to help. They confirm that I should but tell me to head to a regular clinic. They will be closed by the time I can get there. Knowing Misty's distress, I can't be sure she will take anything by mouth, but I wrap the pill in a bit of dog food. It is small, and I expect refusal, but she seems to understand the urgency. She takes it and swallows, gagging and choking wetly before the swallow. When she looks up, I see her mouth has turned a bluish-gray.

The other dogs are concerned, moving in and out like an anxious entourage. Smart too. They give way to our scramble, and we are off as quickly as I can get Misty and my pillow out of the house and down the steps. She doesn't struggle. In the car, I secure her seat-belt harness, tilt her up on the pillow, and am relieved to see her panting mouth pinken slightly as she sits propped upright, though her breathing still seems to bubble in her throat. We head out on the sleepy streets of my small town, with its four-way stops and kindly you-go-first hand motions, and into early-rush-hour freeway traffic on a Friday. It's the tortoise-and-hare commute I've made every day for five months, but now I'm alternately cursing every left-lane driver on a cell phone and pleading with Misty to hang on, as though anything I say to her can help her body as well as her frightened little spirit. I'm watching the road, but I can feel her eyes on me and hear her shift awkwardly on the pillow, tilting left and right in her struggle for air.

Traffic inbound to the city is slow, slower, slowest, and stopped. Something is clearly wrong ahead. As I exit to get onto an access road, Misty gags and falls sideways, shuddering against the seat belt, her open mouth blue and her eyes rolling upward. Clear of traffic now, I veer onto the shoulder, stop, bang on the hazard lights with my knuckle, and take her out of the seat belt and up into my arms, holding her in a sitting position against my chest, ready to breathe into her mouth. She goes limp and her urine soaks through the front of my pajamas, and I think that maybe nothing I can do will be enough to save her. But when

I tilt her a little more in my arms, she gasps deeply and blinks, licking her lips and retracting her blue tongue.

There is no help for it; I have to hold her all the way to the vet like this.

It is a slow drive on the access road, charged and surreal, my own heartbeat thudding dully in my ears. We arrive at the clinic just as they open. I bump open the door with my hip as soon as a receptionist unlocks it. She takes one look at my little blue dog and soaked pajamas and calls a vet tech from where we stand. Without introduction or apology, the vet tech scoops Misty out of my arms and heads for oxygen.

The interim is a blur of paperwork and tedious personal details, like remembering my own name and address. Misty's information comes more easily: her age and general state of health, the historic carpet-tack surgery, the exact date of the congestive heart failure diagnosis, her medications. Filling out those forms is a frustration and a comfort while we wait for the vet's arrival. My hand shakes across the paper, and I feel every sense stretching toward the room where little Misty dog lies. I can hear the tech talking to the vet on the phone quite clearly, giving Misty's particulars in the dispassionate way of professionals, confirming what he's done, confirming what he needs to do further before the vet arrives.

"She looks a little better," I hear him say through the door, and my heart beats time to that awhile. It's the only good word I have.

Twenty minutes later, the vet has arrived, the forms have been completed, and I'm studying a poster promoting dental hygiene and feline health. The vet comes out of Misty's room. He looks young and pink and sleepy. But when we speak, and he pulls out a pad to draw me a diagram of Misty's condition as he sees it in real time, he is calm and competent. This is a bad event, he tells me, but her color has improved under the oxygen, and more Lasix seems to have helped. She is breathing better and sounds stronger, and, when I pick her up this afternoon or tomorrow morning, would I be willing to add another med to the rotation of pills she gets already?

I agree to anything. I agree to everything, elated by the news that she has taken a turn for the better. The vet gives me a high-five (not to tempt the devil, he says as he does so) — my emergency rush for help was right on. Misty wouldn't have made it much longer.

So go, he says, and I'll give you a call this afternoon to tell you whether you should come get her tonight or in the morning. I head home to shower and dress for a day's worth of meetings and a radio interview. I'll have to rush not to be late. But my mind is full of the little Pomeranian. I'll buy Misty a special cushion as a welcome home.

When my cell phone pips ninety minutes later with a text — CALL VET — I know it's not a good thing. When the vet calls sixty seconds after that, it's to tell me Misty is gone.

I can't remember a thing about the two days that follow.

"Not your fault," says Paula, in the early morning of the third.

*Not your fault,* posts a good friend online shortly after that.

"Not your fault," says a search-and-rescue colleague, says pretty much everyone who hears the story of our double loss — Sam the first week of August, and Misty the second. I know my friends mean well, and I know that I did everything I could, but the ghosts of lost minutes and different decisions will follow me for a while, as though a better, smarter Susannah would have wakened earlier or moved faster or done some miraculous other thing that would have made a difference to either of them. I cannot shake the notion that I failed them. It is a consuming grief. Those first days I too have the sensation of drowning.

*Get on with it,* I say to myself, and I do, somehow. The remaining dogs and I follow the greased grooves of our habits together, but the absence of Misty and Sam has affected them too, and I notice they seem in a kind of perpetual hover as we move through the house, their anxious eyes on me as we avoid the spaces the lost dogs once loved.

It is normal to grieve but pointless to brood, I tell myself. *Get on with it,* I repeat like a punishment, *get on with it.* I clean furiously, and I garden with a vengeance. I do more psych dog research. I draft the outline of a novel. I wash all the dogs in turn, whether they need it or not. (I wash Mr. Sprits'l twice in the same day — first and last — and he submits with merely a raise of eyebrows before I realize he was the first dog in the soap earlier that morning. Oh, well. He is very, very clean.) We are busy and productive, by God, making the hours count. Nothing about this cold anger with myself makes me think, *Uh-oh*. It takes the immediate return of nightmares, the same dreams of chained dogs abandoned and rescues failed, to show me how much old trouble is rising.

As I catch myself double- and triple-checking the locked back door one morning (*Uh-oh*), I hear a knock at the front. I'm in a suit and heels, ready for a full day of by God, getting on with it, and I hurry down the hall, ready to be furious with some solicitor trying to sell me something just as I'm heading out to work. I am really ready to take all this out on someone.

At the door is a neighbor I've seen around but don't know, holding a wiggling bundle of white.

"We found this little . . ." he says without introduction. "We don't know what to do. The lady next door says you rescue . . . and we're . . . allergic," he adds. He thrusts the bundle into my arms like it's a peanut he very much fears.

The wiggle squeaks. It's a puppy — all jutting hipbones and ribs and trembling. He looks up at me with desperate eyes, and I can see how sunken they are. It's a pit bull terrier puppy, and he is starving to death.

# 11

OF COURSE I TAKE THE PUPPY. Of course I say yes, though I'll be late for the meeting I was heading out to, and I retreat into the house, where I'm surrounded by furious Pomeranians and inquisitive search dog Puzzle. I lift the puppy high, folding his little bones up and away from all of them (he could be infectious), but Puzzle makes sense of him anyway, trotting behind me with her head up. I can hear the *huff-huff-huff* of her working his scent. I put him in my bathroom with food, water, and a folded quilt, sit on the floor beside him in my suit and heels. He tucks into the kibble and then curls up without protest, which shows how weak he is. A ten-week-old puppy in health would have been bounding, tugging, checking things out, braiding around my ankles like he was wrapping a Maypole. This one eats like he has never had a meal, then flops down as though the eating wore him out. He falls asleep almost immediately, his chin on the lip of the bowl.

I sit in the meeting full of conflicting thoughts, and the white huddle of puppy on a blanket is present in all of them. I am anxious for him, angry that someone allowed him to starve to that state, and I'm dizzy with the sudden logistics of vet trips and — if he survives — having a potentially large-breed puppy in my house full of small, senior dogs.

Not all of my thoughts are selfless. In fact, it's pretty safe to say most of them aren't. The start of a brand-new academic year, this is no time for a puppy. And this one has arrived when I'm not ready to receive him — grieving that I couldn't save Misty and grieving that I put Sam through a second round of dying, and I am probably also angry that I have not yet made peace with the loss of them and with my own sense of failure. In the space of just a few days, the insomnia, guilt, and nightmares I'd known in 2004 have all returned in a rush — the locked door might not really be locked — and now there is all that *and* a puppy, who will not bring peace. I think my heart is closed to this stranger, too sad to be open to another dog so soon. I have no thought that his unplanned presence might rewrite my depression and invigorate the other dogs, who in their dog way also seem to be grieving lost companions. I am too full of burden to be compassionate or wise.

He'll have to go somewhere else. That's all there is to it. Decision made. Firm.

This is something worse than starvation, the vet tells me as he palpates the puppy's abdomen and shakes his head over the thermometer. Less than twenty-four hours after little-mister-no-name arrived, I found blood on his back end and a red bubble at the corner of his mouth, and within an hour, we were at the vet. We made the same trip I'd made with Misty days before. The pup sat in the same seat, wearing the same seat-belt harness, and as we threaded the same anxious way southward, we probably passed the same cars. If I believed in cosmic jokes, I'd be looking for the punch line right about now, I said to the little dog.

As it is, I'm looking down on a veterinarian's metal table at a diminishing puppy that is far too ill to be frightened, too weak to even struggle against the pokes and sticks and swabs, and I'm aware that I'm already hopelessly attached. I've got to find him a place to go — I am still telling myself that — but he deserves good care. Decision made. Firm. "Whatever he needs," I say in the same way I agreed to anything that might have helped Misty, and something lurches inside, because this all feels too familiar, and this little guy looks far worse than she did. After a long scrutiny, the vet can give me a maybe at best. He says the puppy will need days of hospitalization, massive antibiotics, and fluids by IV. Despite all this, he still may not survive.

The vet gives us a private moment before I leave. It is the same quiet move always made at the point of euthanasia, and I wonder if this good doctor, who has known me and all my animals a long while, is telling me indirectly to say goodbye. Almost literally fading before my eyes, the puppy is now too weak to sit up. I kneel down beside the metal table, and he lays his head on my open hand. He is ugly from neglect, patchy fur stretched taut over visible bones, but he is beautiful, nonetheless, all dark eyes and wonky crashed-kite ears. Despite his recent hardship, there is trust in his eyes.

"You stay strong," I tell him. I can hardly speak. I lift his forepaw to my mouth and kiss the spotted pads.

I foolishly hit PetSmart on the way home, as if buying him things, believing in him, gives him a better chance to make it. By evening he has a puppy bed, a collar, and an unreasonable number of toys. He also has a name: Piper, because his ears look like the gull-winged version of that airplane, and Jake, because it's strong. Jake Piper sounds like the detective hero of an old noir film.

At home, the dogs are surely aware of Jake Piper's absence. They sniff and circle all the places where he had been, even though I was on my hands and knees scouring and disinfecting every surface he'd touched before they ever got near. He is there in some way for all of us. I wash all his brand-new, waiting toys. The dogs stand in the doorway of the laundry room and watch me dump the bright lot of fuzzy dinosaurs and stuffed hamburgers, the googly-eyed carrot and the *I am loved* embroidered bear, the whole lot of them into soap and more soap, temperature set on hot.

So much change in a week. None of it escapes them. And whether answering my obvious anxiety or comforting their own, I'm not sure, but all the dogs are following me. Normally, they'd be napping in the middle of a summer day, but now Puzzle trots through the house at my right, her expression thoughtful, and Sprits'l leads the rest of the pack at my left. When I stop, they stop. The Pomeranians' eyebrows raise. Puzzle waits with her tail swaying expectantly. The house is clean. The puppy is gone. *What's next?* the dogs seem to wonder, charged with my restlessness.

The vet said he would call at the end of the day with an update. That he hasn't called already is a good sign, but I'm too wired to sit and wait. What's next, I think, is to find out what went on with Jake Piper before. Maybe someone lost him. Maybe someone is missing him. Maybe someone's *child* is missing him. Maybe he got into this shape by digging under a fence or bolting out a door, and it's no one's fault really, the starvation and infection, and maybe there is as much love on the departure side of Jake Piper as there is, already, where he has arrived with me. (I'm aware that finding whoever might be missing him will likely mean the end of us, a thought I avoid.)

Knowing the day he arrived and from what direction, I clip on Puzzle's lead and out we head into the neighborhood. My little dogs, of course, have something to say about that. They follow our passage as long as they can see us, moving northward through the house along the windows, yapping in protest.

Mine are not the only dogs interested in Puzzle and me. Most of our canine neighbors rush the fence line to greet us. Some announce their territory. Others petition for play. I see the perked ears and wags. I see them stamp with excitement and bow. A few whine to Puzzle, particularly the retrievers. Puzzle whines back, and whatever they are saying to each other seems to make them both happy. I imagine those scenes in *101 Dalmatians* where all the dogs of a city transmit news across neighborhoods to one another, and I wish it were possible to ask Puzzle to ask all those dogs that press their faces toward us if they have any information on a white puppy, about this tall, with crazy ears like this, mostly bones and bad luck, who came from the north.

As it is, we have nothing obvious — no Lost or Missing signs, no anxious person walking the same streets that we are, puppy collar in hand, calling an unfamiliar name. We are a neighborhood with animals, but strangely, everyone with a dog, cat, parrot, or potbellied pig seems to know where his pets are. Apparently, no one has lost a puppy. Of course, Jake's disappearance may have been welcomed by someone. He could well have been dumped at the edge of our dog-loving neighborhood, dumped hungry, which at least was a notch above tossing him into a lake in a bag with a brick. In this town, there are always more puppies than homes.

For a moment a sign tacked up on a telephone pole misleads me. STOLEN, I can see, and WHITE, and my heart flip-flops, but when Puzzle and I get close enough to read the whole sign, I see it's not about a missing dog at all. Someone is missing a plant stand, freshly painted white. Someone thinks that maybe someone else accidentally took it from its painting place, thinking it was a curbside castoff. But it's not! It's a family heirloom! Not worth much to anyone else, but precious to someone's family! Someone would be thrilled to get the plant stand back, no questions asked of someone else. Someone is offering a small reward! It's more an essay than a sign, peppered with exclamation marks, and only a driver going two miles an hour and wearing Mr. Magoo spectacles could read it, but the way it's written and the direction it's facing makes me think that maybe someone has a good idea who took the plant stand, and the essay is an accusation.

"Can I help you?" asks a man loading tools into a truck, a utilities repairman that Puzzle has been flirting with from a distance while I paused uncertainly over the sign. I could feel her simper and wag at the end of the leash.

"I have a little stray puppy at the house and was looking for Lost signs," I answer.

"That's not a lost-dog sign. That's about a missing plant stand."

"I know."

The man wordlessly jerks his chin over his shoulder to a yard full of plant stands of every shape and size. They are in neat, regular rows like a cemetery, all of them white and all of them plantless, but apparently accounted for and loved. He shakes his head slightly, then asks, "What's the dog look like?"

"White. About this big. Large eyes and very long legs. Huge tilted ears."

"I might know that dog," the man says, but he looks doubtful. A white puppy he encountered last week was surely too far away. He recites the bullet points: small, skinny, white, a sick puppy tied in a shed to keep him away from a moving family's other dogs. The family, he says, had been evicted from their home. When they left, they left the puppy too. The repairman heard whimpering and peered into the shed and saw a white scrap of little dog. He called animal services, but they arrived to find an empty shed and what appeared to be frayed clothesline used to tie him down. Either someone had come back for the dog or he'd chewed through it to make his escape. The repairman's description certainly sounds like Jake Piper. But this was blocks away, he says, pointing. He names the cross street. I'm doubtful too. It's a long, long way for a puppy to walk, a lot of streets to cross without getting hit by a car.

If I'm going to trespass, I'm going to do it without Puzzle. I take her home before heading back in the direction the utility man had pointed.

Puz is too light and too visible. I don't anticipate a confrontation, but if there is one, I don't want her caught in the middle.

I take the car and turn left and then right down an unfamiliar street, watching my neighborhood dissolve and give way to another, which melts into another, and then another, Queen Anne architecture yielding to Craftsman and then to blocks of midcentury houses in steadily deteriorating shape. I drive slowly, not exactly sure where the right cross street is, and at the point I think I've overshot the house and shed somehow and am about to turn around, I see it: a once pretty 1920s cottage with peeling paint, sagging porch, and a freshly boarded back door. I can see it clearly from the corner, though it's in the middle of the block, one of a huddle of derelict houses not yet demolished, as others around them must have been. There are several vacant lots for sale.

So that's the house, and there's the shed, much as the utilities guy had described it — as old as the house and well back from it, at the end of a grass-choked gravel drive. There must have been shade over that shed once, but now it abuts an empty lot where both house and trees were bulldozed, and only a few scraggly young hackberries remain. The makeshift crossbar that had once secured the shed's double doors is on the ground, and one of the doors has been pulled open about half a foot. Pulled open with difficulty, it looks like. There's a raw place where the bottom of the door scraped dirt and grass and gravel.

There are No Dumping and No Trespassing signs on most of the empty lots. I park the car and figure that if I walk onto this land as though I own it, maybe someone will think I do — or that I am interested in purchasing it. I sidestep the house and all the shattered glass surrounding it (not sure what has gone on there), moving straight to the shed. The open door is stuck fast in the uneven ground. If I wanted to, I could probably squeeze sideways into the shed through the door's partial opening, but something about that idea raises the hair on the back of my neck. Too easy to get caught and cornered inside. I could see how simple it would be to stick a puppy out here and let it die, unheeded. Out of sight, out of heart, out of hearing. With empty lots on three sides all the to way the corner, there wouldn't be a whole lot of people who could hear a little dog cry.

Peering through the gloom, staring straight into sunlight pouring through the generous cracks in some of the walls, I find the shed's interior difficult to see, but to one side I can make out the metal skeleton of an old lawn chair, a gas can, a bike missing a front wheel, and a shredded garbage bag full of something. That's it. The squeeze is too tight to turn my head.

No good.

I'm perhaps the most incompetent trespasser ever, uneasy, furtive, glancing around to see if anyone's watching me. No one. I shove my hands in the pockets of my khakis and circle the shed, flustering insects in the tall weeds. An angry grasshopper levitates, and all sorts of flying things rise up in a cloud around me as I pass. On the ground, more broken glass. A headless doll and a crumpled Big Gulp cup. A screwdriver with a black-and-yellow handle protruding from the dirt. I am looking for any way to see inside the shed when I find a low hole in the broken plywood at my feet. The wood is rotted. When I kneel to the hole to peer through it, the wood crumbles easily to the touch.

Face pressed to the opening, I smell the inside of the shed before I can clearly see into it. The familiar, sickening stench of animal waste and, behind it, mildew and something very like the sweet-foul scent of decomposition meet me at once, and I reflexively pull back. I have smelled animal death in a hoarder's house and a puppy mill, and I've smelled plenty of animal death in other places. Once you know the scent of that despair, it never goes away. For a moment I can't look, afraid that there may have been a litter of puppies in this ugly story, and that only one escaped.

I inhale and hold it, again put my face to the shed to see what I can see in the space of a breath. There's the bike and the lawn chair. There's the garbage bag and there, beside it, two very large, very dead rats. There's another — a third stiff rat on the bag itself. Near the back of the shed, a pair of plastic mixing bowls are overturned on the dusty ground. A ragged end of clothesline droops from where it is tied to a metal toolbox not far away.

That's all I can see. It's enough. The shed tallies with the repairman's description, and if a little white dog had been in there, that little white dog is gone — mercifully gone, maybe — and not lying among the dead.

Was this Jake Piper's beginning? How many white puppies in trouble could there be in the space of a week? And if it was Jake in this garage, how did he avoid whatever killed those rats? How did he manage to make his way to me?

I push back, get up, and turn away from the shed, wondering how many times animal control officers walk up to places like this and prepare themselves for the worst. I wonder how many times they find it. It is a job I could never, ever do.

# 12

WAS THERE EVER ANY DOUBT that Jake was staying? Probably not. *At least for a little while,* I rationalize about the collar and the toys from PetSmart. I posture left and right about finding him a home, convincing no one. Even had I been more determined that he needed to go, talking to fellow rescuers would have changed my mind. They are unanimous: There are too many homeless puppies these days. Most rescue shelters were overloaded and would find a sick pit bull mix unadoptable. This kind of puppy wouldn't last a day in a shelter, several tell me.

Certainly my vet knew Jake would stay where he landed. For four days there are grave reports and guarded prognoses, but on the fifth day he calls to say that "Jake Piper is up this morning and playing like a puppy. He can go home tomorrow." Then he asks if I want him microchipped.

I do. I send the papers off in my name, because, really, he has to be registered to *someone.*

Now as we sit together on the floor, I can see how much he's improved and how far he has to go. Jake is still terribly thin, but he's quietly playful, resting in my lap and gnawing at the long tails of toys I dangle over him. The Poms aren't certain what to make of him, this puppy that, even skeletal and weak, is already bigger than they are. Perhaps they sense his frailty. Often a new rescue in the house will rev them up for an evening, but not this time. They sit in the doorway and watch speculatively. Jake is still so dazed he hardly seems to know the Poms are there.

But that's not the case with the golden retriever. The youngster is devoted to Puzzle. When she passes by him tonight, he stretches out skinny legs to tap her with his paws, a sort of doggy Morse. *Hey. Stop. Hey. Me! Notice me!* Puzzle was spayed before she ever had a litter, but her maternal instinct has kicked in. This surprises me. She alternately cradles and corrects Jake as though he were her own. Now she stops in passing and gives his ears an affectionate lick.

Jake will be slow to regain strength. The vet has made it clear just how close we were to losing him. Starvation, hookworms, and hemorrhagic gastroenteritis. Ugly words all, and they have all taken a toll. He is a ghost puppy these first weeks back at our house — all knobby joints, tucked tail, and big dark eyes. His strength is elusive. "He has turned a corner" is what they say about an invalid who's condition has improved, but sometimes, literally turning a corner is too much for Jake. There are days he walks down the hall like an old dog. I watch his legs buckle, or I hear the clatter of his bones on the hardwood floor and look to find him lying on his chest, bemused and panting, gathering strength to make it to the water dish.

The Poms follow him from a distance like a little Greek chorus, commenting on his successes and his failures. Those familiar with Pomeranians could probably chart Jake Piper's progress across the changes in their behavior — from subdued muttering to spins with chittering to insulted yapping when Jake crosses their invisible boundary lines. As he improves, he gets bigger and stronger. Now that they've figured him out, the Poms, unimpressed, draw more lines. They follow him more closely. Correct him more often. Not that couch, that cushion, bed, or bowl. He complies.

A friend sees a photo and comments that Jake Piper "is looking a little less like Gollum and a little more like Yoda these days." Her description is right on. Certainly Jake has Yoda's ears. What to make of them? Not floppy and not upstanding, they are enormous sails and unevenly folded. Wayward too. When Jake Piper walks south, one ear heads that direction and the other points east. The southbound ear has rusty spots rimmed in dark gray. The eastbound ear has rust polka dots. The same friend who called him Yoda describes the ear mottling as "a cross between mold and mildew." A few observers say he's adorable; a handful call him cute, stressing the word in the way they'd use it to refer to ugly babies — falsely bright — then change the subject.

One search-and-rescue colleague cuts right to it and says that what Jake lacks in beauty, he may make up for in brains.

He's got a good nose, certainly. I find this out early, when he scents a single piece of dropped kibble through a closed cupboard door.

Because he's still shy of full strength, it's difficult to know what kind of dog Jake Piper's going to be. I watch for hints about his drives and aptitudes. Like Puzzle, he's a nose-y beast. The way one dog will model another, Jake mimics her working scent, but I can already tell that he's far less independent than my search dog. Jake Piper is a softhearted boy. He's curious but friendly with cats and submissive to even the smallest of the Pomeranians.

Jake likes a toy. He likes a ball a lot, with an almost Border collie obsession for fetch. But bottom line, Jake prefers other dogs to toys, and he prefers humans to dogs. Jake loves a human most — and not just those he knows. All humans. Any humans. Especially children, as he proves when I test him on a leash. Neighbor children are charmed by Jake's long-legged, kite-eared appearance and his large, dark eyes, and when they approach and ask to pet him, he is gentle and eager. While he shows no sign of separation anxiety when I leave for the day, Jake is not Puzzle, who enjoys human contact on her own terms, who prefers her work off-lead and yards and yards away. No: Like Misty, Jake glows with human contact. He wants his people close.

I wonder about Jake Piper as a working partner. I deliberate in e-mails to friends with working dogs. While I won't test him yet, I won't force him, and Jake has no obligation to be anything more than a good family pet, he is so unusual a puppy that I'm curious.

One night, in the middle of Jake's wrestle game with Puzzle, I touch a finger to my cheek. "Jake, look at me," I say. It's the first time we've tried this. Jake knows his name. He turns his attention from Puzzle and looks into my face.

"Good look!" I praise him. "*Good look.*"

And then Jake turns upside down in front of Puzzle, wraps his paws around his nose, and snorts. He peeks up at me, keeping the eye contact while also showing off, and for the first time since Sam and Misty died, I feel my sadness lift.

•   •   •

A rescued dog's ability to give up old baggage often tells us what kind of dog he can be, whether pet or working partner. Starving Jake came with plenty of baggage. He is never far away from that shed and the clothesline that bound him. In many ways, it seems he's never left it.

Food is his greatest worry. I have had plenty of starved dogs in the house before, but Jake's desperation exceeds any dog's in my past experience. Jake is hungry before meals, during meals, and after meals. He is so hungry that sometimes he whimpers as he eats, a jumble of impulses, unable to recognize that the anxious croon for dinner is being satisfied even as he moans for it. Jake is hungry even when he really can't be. He wolfs down food like it may vanish at any moment. This is understandable, of course, but his early manners with us are atrocious. He often knocks the food bowl from my hand before I can put it down.

It's more than an issue of manners. In time, this frantic eating could be dangerous for Jake Piper. Bloat — a horrible and often fatal condition that can result from a dog's gulping meals, swallowing air, and then doing some simple activity afterward — is especially possible in deep-chested dogs like Jake. I need to teach him to slow down. I do a standard trick to slow a gulping dog: I put his food into a muffin tin so that he has to extract it with his tongue before he can chew and swallow, and I slow him down even further by greasing the bottom of the tin with peanut butter. Jake is still faster than every other dog in the house. He punctures several muffin tins with sharp, eager teeth when the meal is finished, still kibbling even though the kibble is gone. When I pull the tin out of his crate, it looks like it's been stabbed with an ice pick.

So Jake has a lot to learn. I have compassion for his history and respect for his intelligence and his willing heart. I believe that a slower, more civilized dinnertime Jake is possible, and it will be a huge indicator of his willingness to trust humans for good things. Even though he's desperately overeager for food, and even though he's hell on a muffin tin, Jake is also full of surprises here. He has known starvation, but Jake doesn't guard his food with humans or other dogs. I can touch his bowl at any point while he's eating without fear. Another dog can even approach his crate at mealtimes. While Jake doesn't offer to *share* with the others, he also doesn't give the sidelong stare or growl, and he doesn't bare his teeth. I have a golden retriever who's never been denied a meal in her life that didn't come into the family that gracious.

That softness from a dog that has starved, that early trust, is a very good sign from Jake.

We begin training good manners with Sit, one of the single best contracts between a dog and a human. A dog that sits necessarily gives up power. I still remember a favorite dog trainer's mantra: "Sit for petting. Sit for treats. Sit because I told you to sit. The dog should think his middle name is Sit." For a dog like Jake, who has no reason to have faith in humans, who had to fight his way out of a shed just to stay alive, the test of Sit is the first real look at just what his background has done to him and just what kind of dog he is willing to become.

Jake is attentive and he thoroughly models Puzzle. He learns the Sit command quickly after only a few tries. Competitive Jake seems to want to best Puzzle at being good. Jake enjoys being good. He sits for toys quickly. He sits for petting, too, but is sometimes so overcome with the joy of a fine scratching that he melts, leaning against my legs and sliding belly-up onto the floor. Sitting for treats is harder. Jake sits for them and then pops up the minute a treat is given. Food is such a joyful thing. It takes a few days for him to keep himself planted on the floor throughout the whole process: sit, treat, chew, swallow.

Then we begin working on mealtime etiquette. "Kennel Up!" is the next command, and Jake learns quickly that he should run to his crate, sit in it, and wait for his food. But God help any dog or cat that gets in the way. Jake makes the quick connection of command to the kennel, but sometimes his enthusiasm causes him to pogo during the gallop down the hall. I fall over him a few times. He hits the bowl with his head a few times too, the kibble scattering like shrapnel causing a frenzy of Pomeranians and an eager Jake attempting to shoehorn himself under bookcases for a single piece of food.

Jake is determined and enterprising. He will go for that stray piece of kibble. Once I find only his hindquarters and tail sticking out from under a three-hundred-year-old writing desk that's topped by a high glass bookcase. This secretary has teetered before without provocation, but it is oddly stable now. I hold on to it and watch Jake slither smoothly out from beneath, crunching a single piece of dog food, his froglike back legs pressed to the floor as though he were hipless. I know that mice can do this sort of thing, but how does the dog make himself that flat?

Clearly Kennel Up needs augmentation. I'd like to say, *Stop @#$%# hopping*, but I'm not sure Jake would understand it. So instead I stop walking, food bowl in hand, every time he hops. The first time I do this, Jake is clearly bewildered. He hops a little higher, a little faster — boing, boing, boing, *boing* — every hop a prompt to get me going again. (I realize then that Jake's frantic hopping has made me walk a little faster with the bowl, but now when his hopping makes me stop, I have undone all the careful training he has given me.) For a couple of days he tests my will in this, scrambling, spinning, hopping straight up like he is catching a Frisbee, shooting me a dirty look every time I stop. And behind the door in the laundry room, a rabble of angry Poms are impatient too. They bark: *Hey, us! Hey, food! Hey, us! Hey, food! Hey-hey-hey!*

It would be so easy to give in.

But on the third day, God created understanding. And obedience. I pick up Jake's bowl and he trots to his crate and sits inside while I rest his bowl on a bookcase and feed two other dogs in their crates on the way. We are officially hop-free. Jake's self-discipline is coming along. Now to explore what else he wants to offer.

Jake Piper has the gift of gaze, and he's looking at me from the end of our long hallway. He is a ghostly little figure on this rainy morning in November, sitting in his good-boy posture, bathed in the soft light through the tall windows on either side of our old house. He's still thinner than he should be, but the harsh outlines of his ribs have faded, and his hipbones no longer jut out. Still, even Jake's best friends might call him funny-looking. From here, I can see the several breeds wrestling across his appearance, this boy with coltish legs and misdirected ears and curious, bendy tail, but his gaze is steady and thoughtful, his intelligence obvious. Jake has beautiful eyes, dark almonds rimmed with black.

"Guyliner," a friend calls this look. A little bit goth. A little bit rock-and-roll. A little bit Winona Ryder, if she were a small, spotted dog.

He has a lot of patience for a youngster, but from down the hallway, he's urging me with those electric eyes. He's learned the Sit and Stay commands, and I have put him in that Sit, walked to the end of the house, and stood very near the treat drawer. I can see that Jake's every muscle wants him to get up and force the issue. But he holds his Sit, head tilting this way and that while he waits for the word of release. In a farther room, the Poms are protesting because they aren't in on the action, whatever it is. Two cats watch from my bedroom doorway. They look at Jake with their ears swung back, speculative. They like to watch this training stuff from a distance. For them, newcomer Jake is still a guided missile likely to go off the grid.

He watches and waits. I touch the treat drawer, and I see him lick his lips, legs quivering.

"Jakey, come!" I say, and he springs forward so strongly that the cats watching from doorways scramble, their backs arched and tails bushed out. Jake is still unpredictable. He clatters down the hall-way, puts on the brakes in the kitchen too late, backpedals a little like Elmer Fudd, and slams into the cabinet door beneath the treat drawer. Unfazed, after the slide and the bang, he looks up at me with a loopy grin, adopting a second Sit required for the treat. Those dark eyes are bright. A thread of drool slips out of his mouth and onto the floor. He gets a Milk-Bone and a homemade peanut butter kiss and a *Good-boy-oh-what-a-good-boy-you-are-so-very-good-Jake-Piper!*

Jake capers a little when he finishes his treats, then drops to a third Sit, looking up at me fondly. I kneel down to him, holding his gaze. Treats are great rewards for Jake, but in its way, bound up with pride, affection, and reward, human connection is even better.

"You're a good boy, aren't you, Jakey P?" I murmur.

He leans in, looks tenderly into my eyes, and burps.

And then there is the opinion of Maddye, the cat. Maddye is a senior tortoiseshell, a most ancient crone approaching sixteen. She is one of two cats here. She is older than every other animal in the house, and, a rescue herself, she has seen the other rescues come and go — a whole host of foster dogs that either went to other homes or lived with us until they died. She has been instrumental in raising and correcting them all.

Now she is not sure what to make of Jake Piper. When he first showed up, Maddye was nonplussed. How to understand a puppy that mostly slept and wobbled about the house like he was very, very old? For Maddye, he was a troubling oddity. He had a habit of napping wherever he landed when his energy failed. Often he flopped down in Maddye's favorite sunning places or in the doorways of rooms she liked to haunt.

Early on, she avoided him altogether. I tried to imagine her cat sense of him: the strange appearance, the smell of illness and of vet clinic. But Maddye was certainly full of her own cat curiosity, and in time, she took to shadowing him from a distance. She would creep into rooms where he was sleeping and glare. Sometimes age got the better of her. More than once, I saw her stare melt into general drowsiness, and the old cat would fall asleep in her stalking place, her nose drooping to rest on the floor.

Later, she grew bolder. She crept close enough to sniff. If sleeping Jake twitched in a dream, back she'd jump, landing on light feet, then start the approach all over again. Maddye had an agenda. Jake's disproportionate ears seemed to have a life of their own, waving gently beneath the ceiling fan. The cat was fascinated by those ears. Once she couldn't resist the lightest touch with a paw, and Jake groaned in his sleep, which set Maddye back a few feet, wide-eyed. She would try again later. She got away with the stolen ear-tap a few times, and then she was bored with getting away with it. I'm convinced the old cat was ready to start something. One late evening, Maddye decided to *bring it*. I saw the splendid cat moment of poise, her paw raised and slightly curled for a long moment before — *wham! wham-wham-wham-wham-wham!* — she smacked sleeping Jake's ear hard.

Jake made a cartoon *yoik*, and Maddye dashed off with a screech — *It is alive! Alive!* — Jake scrambling after her and barking at the top of his lungs. It was the first bark I had ever heard from the puppy. He didn't catch her, but a line had been drawn.

Now Maddye frequently slaps Jake awake. Jake, in turn, has begun to seek Maddye out during her own naps. She likes to tuck up in precarious places to catch the sun — the arms of chairs, the edge of desktops, narrow windowsills. I often wonder how she doesn't fall. Maddye's naps grow longer as she ages, and I can't always find her when I come into the house, but Jake can. More than once, he startles Maddye awake, leaping into the chair or bouncing up to hit the desk with his forepaws. She always scrambles. Sometimes she spits. It's a most satisfying result for Jake, whose good-natured harassment never crosses the line into real threat.

But as he gets taller and stronger, and I'm aware of his increasing prey drive, I begin to discourage Jake from taunting Maddye. Cat-friendliness is a requirement for dogs in this house, and while I feel sure he wouldn't hurt her intentionally, he's big enough now that he might.

"Jake, Leave the Kitty!" I say when I catch him about to flush her out of her sleep spot. He looks surprised, but he stops. After a few rounds of the Leave the Kitty command, he stops hassling Maddye when I say it, though there's real disappointment on his face.

There's intelligence too.

One afternoon I'm in the kitchen prepping vegetables for a salad, and I hear the *tip-tip-tip, huff-huff-huff* that says Jake is behind me working scent. It sounds very much like he has found the cat. I turn and spot Maddye tucked up on the edge of a chair, head bowed to the upholstered seat. She is snoring. Jake has tiptoed up to stand very still about six inches from her sleeping place. One snort, one bark, one paw on the edge of the chair, would startle her awake. Moreover, based on where she's sleeping, she's likely to go straight over the side of the chair and into the water bowl beneath. Though I don't think Jake processes all that, he would no doubt like to dunk the cat.

"Jake . . ." I'm about to caution him when the dog looks up at me, looks at Maddye, and then looks up at me again, his expression transparent. *Give me a reason,* he seems to say. *Give me one . . . good . . . reason . . . why I shouldn't.*

Theorists will tell you background and breed traits are important qualifiers for working dogs. Some theorists go on to say that the breed that dominates in a mixed-breed dog's genetic makeup may be a strong predictor of the dog's success at work. This could be interesting. What if through some doggy liaison Jake Piper looks something like a shepherd but acts more like a Shih Tzu?

Friends are speculative. They're pretty sure they see German shepherd, pit bull terrier, maybe even a little Labrador retriever. Will Jake's unknown mix result in a fabulous dog willing to learn anything or a conflicted dog with disastrous impulses? The good news: Jake seems to come from working breeds. The bad news: Jake is unlikely ever to win a beauty pageant.

We move forward with basic good-dog training and an eye to Jake's aptitudes. Now that he's healthy enough to play without getting winded and to come upstairs without collapsing, it's time to see if Jake wants to be a working dog. Just what would he most likely enjoy doing? Search-and-rescue? Arson detection? Therapy? Service? Of course, there's a chance Jake won't be suited for work at all. Which is fine. Anyone who's come home to an overjoyed dog knows that love can be more than enough.

# 13

HER HAND WOULD SHUSH the alarm clock before it made a second ping, like a fledgling being sat on by a mother bird wary of cats. She had a pathological fear of being late. Merion tells me it was ridiculous, that constant awareness, even dread, of time, but in her worst days she had actually been afraid of what might happen if she rolled over a little more slowly, put her feet to the floor a half a beat after she should. To do so would open the door to all kinds of trouble. She never let time get away. Her routine through the house was so tightly clocked and narrowly grooved that she actually wore paths in the carpet. Her daughters shook their heads and pointed it out; her youngest grandson had chugged his feet along the track between living room, bedroom, and kitchen making *choo-choo* sounds last Christmas. They all laughed about it, but "Jesus," Merion says, "it would have been good not to carry the weight of time on my back everywhere, like it was the only truth, like I alone was responsible for the pulse of the world."

She is tall and elegant and beautifully composed. She is also a widow. She has been alone for several years. A longtime educator, with her punctuality, neat handwriting, and due diligence in all corners, she successfully hid the whole mess of herself from everyone for a good long while. Correction: She thought she was hiding the whole mess of herself from everyone, until two of the newer teachers in her school made complaints about her that fell somewhere between "creating a hostile work environment" and "harassment."

Merion has always kept an eye on things. She walks the halls of her school. A few years ago, she began looking into classrooms a lot. (Maybe too much.) Unused to these supervisory rituals, the two new teachers resented the once-an-hour routine of her opening the door to their rooms, peering in, taking a student head count, and shutting the door again. Merion's habit took root after the Columbine shooting. The older teachers never seemed to have a problem with it (they were all jumpy then), so they didn't complain even when their head's behavior escalated from an every-two-hour check to a once-an-hour walkthrough. Even the students mostly ignored her. But more recent students seemed to feel these new teachers' angry tension, and Merion sensed it wouldn't be long until "classroom disruption" was added to the list of complaints on her annual evaluation. The very idea made her feel sick.

She knew she was overdoing it. She didn't know how to stop.

These young teachers were newly schooled and terribly current on their rights. They seemed to Merion bristly with antennae, waiting for some slight or harm. It was all very procedural. They spoke between themselves and then spoke privately to her. It was made clear: *When you keep checking on the classroom, we feel as though you don't trust us.* She felt her gut tightening in the presence of that deferential first challenge. She answered them more harshly than she meant to. Her school, her rules. There had never been any trouble here, and there was a reason for that. Merion says she knows how to be more collegial — she could have given excuses that mollified them or made them laugh; she could have made some attempt to back down to the old two-hour check she'd done for years; she could simply have told them the truth and said that she was working on it. But she felt herself getting older and more fragile. Her center wavered. It was harder to contain.

And all this time management and double-checking her classrooms took her away from other administrative duties, duties that had to be performed far faster than they used to. Some of her superiors complained that her response time back to them was slower; they expected same-day e-mail responses, and they couldn't reach her by phone. So the long and short of it was she was in trouble at work — stunned to find herself in this position after more than two decades of strong service.

A meeting with one administrator led to a meeting with a counselor, who went on to advise an appointment with a therapist. Merion sat with the thoughtful young woman for two sessions, her guts quivering, mentally ticking off the cost by the minute before she talked her way back to Columbine and then back farther, to the loss of a brother in a drive-by shooting while they waited together for a school bus when she was eight years old. She barely remembered the event, she said to the therapist, but she cried as though she remembered it too well. A screech of tires, voices shouting. *Pop, pop, pop.* Her brother was down. Her brother was gone.

Children killing children with such casual malice.

Merion still doesn't understand how this relates to her stranglehold on time, but she's smart enough to understand that nothing good comes from it. There was catastrophic thinking on every side — a world of imagined things would fall apart if she didn't operate on schedule; a number of equally bad things could happen if she did, one of which was losing a job she needed in a profession she loved. She needed help and feared it. She feared medication most of all.

After several months of sessions, her therapist suggested that an assistance dog might help. She said there were all kinds of things these dogs could do to redirect obsessive thoughts and compulsive behaviors. They worked very well with thoughtful, analytic personalities.

*A dog?* Merion thought. She'd never had a dog. She didn't hate them, but she didn't love them. They could be nice, sure. Smart, maybe. But really, she had no clue about dogs at all. "My God," she said. She remembers looking at her watch and glancing at door. "Don't dogs take a lot of time?"

Ten years ago, six months ago, one day ago, Merion would not have been able to imagine herself here, sitting in an unfamiliar house, upright in her good casual clothes, her back not touching the chair. Trying not to show how scared she was. There was a magazine upside down on a table: "Ten Ways to Be the Best You Ever," she read, or something like that, and the promise seemed ludicrous. She was waiting for the sight of a dog that might change her life.

Merion had been up since dawn preparing herself, which was a little harder when it was possible you might walk out the door empty-handed and come back with a seventy-five-pound stranger ahead of you on a leash. Her never-had-a-dogness felt pretty overwhelming at dawn. Would the creature really foul the floors and pepper the upholstery with fleas? At 7:30 in the morning, she found herself fiddling and fussing with a tag on the bottom of a new water bowl, bought especially for the day when she actually met the right dog. Damn sticky tag should have come right off, but it didn't. She picked and scrubbed at it until the pads of her fingers stung, then stood with her eyes closed, the warm water running over her hands. Then, for some reason she doesn't quite understand, she dried the bowl, set it neatly aside, went into the bathroom, rummaged through her late husband's things, and clipped her hair close. What was this? A real buzzcut. With every pass of the clippers, she felt lifted and naked and new.

Sitting in her nice slacks and houndstooth jacket, Merion could feel wind across her scalp. Last time that had happened, she was a baby.

Why exactly did she cut off her hair?

What would these people think of her?

Did she care?

Merion says now she thinks maybe almost bald suits her. Maybe she looks proud and strong, not like the victim of a fever or the survivor of a war. At the time, however, she wasn't so sure. When her therapist saw her, she said only, "I'm jealous. You've got the cheekbones to pull that off." Merion's mother had the cheekbones, and her brother, from what she remembers of him.

God, she was nervous, thinking about anything in order to take the pressure off the moment. It had been forever since she trusted anyone other than herself, and now she was about to hand it all over to a dog.

A door opened, then closed. Merion heard a jingle of tags and a steady *pad-pad-pad* and someone talking low. Then the dog was in the room. The dog was not that large, but she seemed to fill the space, or maybe Merion pushed everything else aside at the sight of her: a fawn-and-black beast with huge upright ears. The dog stood in the room as though she owned it; she'd been fostered here awhile. The dog lifted her head thoughtfully, flicked an ear: strangers.

"Go ahead," encouraged the foster lady.

The dog first politely acknowledged Merion's therapist, who sat closest, and then Merion, who was visibly shaking when she stretched out her hand. The woman of the house called her a Belgian Tervuren, maybe mixed with something else.

"Hello, Belgian Tervuren," Merion said, or tried to say, the words like a mouthful of fuzzy marbles. She could hear her voice wobble. She was ashamed of her trembling hands but didn't withdraw them, forcing honesty out of herself in front of a stranger and the dog, who sniffed her outstretched palms carefully, then touch-touch-touch-nosed down to her raw fingertips. Merion wonders if the dog could have initially smelled her fear, her sense of inadequacy.

When Merion reached out to stroke the dog, she sat immediately. Startled by the movement, Merion withdrew her hand.

"She's *supposed* to sit for petting," said the foster.

"Wonderful," said Merion, trying again. The dog held her Sit and submitted patiently to the caress, gazing up at Merion with intelligent, kindly eyes. She was magnificent, and Merion was humbled and scared. A dog, for God's sake, but she was in awe.

The dog was from a military family that had had to leave her when they went overseas. A quiet dog. A good dog with kids and other animals because the family had everything — dogs, cats, fish, turtles even. Merion wondered what went on in a dog's heart with that kind of goodbye. The foster lady looked Merion in the eye and said she understood this would be Merion's first dog. Merion might be relieved to know the dog was housebroken. Merion was.

The dog seemed gently fascinated by the leather band of her wristwatch.

The foster lady said: "Housebroken or not, she needs activity and routine exercise, and we liked that she might learn a job to be with you every day. Do you like to walk?"

"I do like to walk. I walk a lot."

Merion's therapist smiled at that.

"Her name is Annalise," said the foster lady, "but the family called her Annie."

The dog's ears perked; she flashed a glance over her shoulder at the foster lady. Was that a call to come? She looked curious and torn.

"Annie," said Merion, leaning toward the dog from where she sat. "Annie!"

Annie looked back, eyebrows raised and ears cupping to the sound of Merion's voice. She opened her mouth to flash out what seemed like a smile. *Huh!* she panted. Merion bowed forward where she sat, and Annie lifted her nose to Merion's shaved scalp, huffed a kiss across it. Merion says she laughed at the sensation, somewhere between a benediction and a prank. It was her first laugh in a long time. She was in over her head, and she liked it.

Foster lady wasn't kidding. This dog loved to walk. Annie had been home only hours, but already her bowels had sent them both outside with pointed urgency, multiple times. Ever the big planner and thinking to spare the yard, Merion had bought dog waste bags in bulk. She'd also mapped out a walk route that terminated in a park trash bin. In these best-laid plans, her fantasy dog would conveniently evacuate just steps away from the bin (perhaps behind it, for modesty's sake), and the majority of the walk would be an idyllic ramble through Merion's aging neighborhood, the overgrown wildflower gardens tilting their blooms obligingly to them both.

But damn. Not this dog, that day.

Merion desperately wanted to have things in control, but she swallowed her pride to call the foster lady six hours later. Annie didn't seem sick, just restless and urgent and mysterious. Was she unhappy? Pining? Was a third home in the course of a year too much?

The foster lady believed Annie could be both excited over the new surroundings and pining for the old.

It was a new thought to Merion, that the dog might be excited about all of this too. Happily, Merion hoped. Could a dog already like her that much?

The foster lady suggested Merion take her on walks, feed her simply, give her time. If she wasn't better in twenty-four hours, contact a vet.

Merion hung up, uneasily reassured. It was time for bed, and she was beyond tired enough to go there, but she thought maybe Annie needed to be close to the door that first night, and she thought maybe she needed to be close to Annie. Merion sat in the big chair that was once her husband's, a saggy, outdated plaid thing that had never been hers. It had been empty since he left her, not sacred, not a shrine, no matter what her daughters thought of her keeping the chair he had died in. Now Merion sat in the hollow her husband had left, propping her feet on the ottoman, just beyond the indents his own feet had made. He had been shorter. *Legs,* he had called her, with such affection. God, she missed that.

Annie settled alongside the chair. The dog's posture was alert, her ears poised. Merion says Annie must have looked like a bat when she was a puppy. She could catch anything with those ears. She could catch everything.

"So here we all are," Merion said to the dog. She picked up the TV remote. It was a foreign object, rarely used.

TV on.

TV off.

Cable on.

TV off.

TV on.

*Bip, bip, bip* — she pressed random buttons. Bert hadn't watched much television until that last year. Merion remembers the lonely sound of his worst hours in this chair, the flash fragments of his channel-surfing that merged CNN with HBO with Discovery and Nick at Nite, a mosaic of half-told stories to distract from the terrible end of his own. Now she was at it. Late-night television — not so much a wasteland as a yard sale. *Bip. Bip. Bip.* Annie's head tilted with every change onscreen. Choose something, Merion. She rested her hands in her lap and looked at the dog. Her dog. They would stay together all night in the intersection of lamplight and shadow, fall asleep watching reruns of *Everybody Loves Raymond.* Annie would finally rest. Merion would forget to set the clock.

That weekend brought interviews with three dog trainers. Merion and Annie traveled to meet the first one. Annie loaded easily into the back seat of the car. She was patient with Merion's awkward fumbling with her seat-belt harness. With her great collie-like ruff of fur and regal demeanor, the dog reminded Merion a bit of some elegant Hollywood star at a fitting for the Oscars. Annie lifted a foreleg graciously to accommodate a strap, giving Merion's knuckle a little kiss in passing. Merion stepped back and studied the dog in the back seat of her car. Annie looked secure enough and safe. She was calm — even, Merion thought, a little subdued. Was this the dog's genuine nature, or was this the dog expecting another displacement?

The first dog trainer was a retired police officer. He had a kennel and a training facility just outside the city limits — a wide ranch house with red shutters beneath a gracious sweep of old trees. Tidy bushes trimmed square. As she pulled into the driveway, Merion could see something like a paddock near cinder-block-and-chainlink kennels deep in the back. The man was prompt, out the front door already wearing sunglasses and walking down the sidewalk as she rolled to a stop. When she got out of the car, he nodded politely but stood where he was with his hands on his hips, watching in a posture that felt, to her, vaguely confrontational.

Or maybe, she says now, she was just touchy.

Merion paused uncertainly.

"Get out the dog," he shouted from fifteen yards away. "I want to see how you handle her."

Merion moved to the back seat of the Sebring, where Annie waited without struggle, lifting her nose to the new environment as Merion released her harness. Merion took up the lead, and Annie lightly jumped from the car. The dog stepped forward toward the trainer and then paused, turning to look at Merion in question.

"Okay!" said the trainer — overloud, in Merion's opinion — and clapped his hands for attention. "First thing. Dog doesn't get out of the car until you tell her to. *Tell her.* She shouldn't just assume. Second thing. Dog doesn't move from her place outside the car until you tell her to. Dog already has way too much control here."

Merion nodded. Annie seemed to feel the tension on the lead and sat.

The trainer called: "So what're you going to do?"

"You mean in the future?"

"I mean *now.*"

Was that sarcasm she was hearing? Merion wondered.

"Dogs are smart. You've got to be the boss. Right here. Right now. Are you cut out to be the boss?"

"You want me to give the dog a series of commands and see her do them."

"*Exactly.*"

He was smirking, the $130-an-hour son of a bitch.

Merion took a deep breath. "Annie," she said quietly, "load up." And with that she gestured the dog into the car. Annie leaped up gracefully. A tug of the seat-belt harness over her head and a snap of two clasps and Annie was secure again. Merion shut the back door.

"Fine," said the trainer with another double clap. "Let's start over."

"Let's not," said Merion. And she was in the car with her head turned toward Annie, backing down the gravel curve of driveway. Merion says now that maybe it was just semantics, and maybe they just got off on the wrong foot with the guy, but the thought of opening herself up to him, the thought of a life riding roughshod over this gentle, wistful creature, made Merion feel sick.

Twelve minutes wasted, plus drive time there and back again.

The second trainer stood them up. No e-mail. No phone call. No nothing. Merion felt a little frustrated and a little relieved. What to do with a lost hour meant to teach them what to do? She remembered an ice cream shop that had a special K9 Cone that supposedly tasted like hamburger. Disgusting, but maybe Annie would like it. She and Annie drove there, and Annie puzzled over the chilly steak-colored concoction, nosing and lapping idly at it, while Merion had a mango milk shake and tried to muster up a few *Oms* from her yoga days. Then Annie looked up. She had a dollop of ice cream across her muzzle, round and red, very like a clown nose. Her tail swished idly, and she leaned in a little when Merion reached forward to wipe it clean.

"Hello, missy," Merion said, rubbing her ears, and the swish accelerated. So small a thing to people who've known dogs all their lives — the kiss and the wag — but to Merion, it was a funny, friendly little revelation. Still is, she says. Still is.

The next day, the third prospective trainer kept them waiting. Merion and Annie got out of the car outside the facility, a former hair studio, all pink stucco and terra-cotta tile. Theirs was the only car in the lot. Meticulous Merion checked — the address was correct. BLOWOUTS and HI-LITES and FADES, Merion read on the exterior wall, and UNISEX, the words still visible beneath a hasty coat of paint. She peeked in the window, cupping her hands around her eyes, and could see the faint shadow of a counter, linoleum tile beneath. If she hadn't known better, she would have sworn this place was vacant. There was nothing dog school about it. There was nothing even vaguely professional. She could see antacid-pink walls, the orange traffic cones spaced evenly over the linoleum. Four orange cones on one side. Three on the other. An obstacle course? Merion wondered. What was this place?

The dog-training facility was part of a duplex shared with a little vet clinic. Merion recalls heading back to the car to get her cell phone when an elderly man with a long silver ponytail wobbled out of the clinic, his arms full of a droopy, Toto-looking dog whose dazed head wobbled too. Once upon a time, Merion wouldn't have given the pair a second glance, but now she was a dog owner, and she felt proud and somehow connected to a fellow dog owner. The man struggled to open the door of his car, where the dog's carrier sat, and Merion moved with Annie to help him.

"Sweet little thing," she said to the gentleman, opening the car door. The old man grinned. "For now. He's still pretty juiced up."

"Has he been sick?"

"No. Nut job."

What does that mean? Merion wondered.

"Little bastard's been humping everything."

"Ah."

"I warned him: *snip, snip, and you're done for.* But did he listen? No."

"Well."

The sun was warm through the cool air. The old man put the dog in his carrier and turned his face to the sun, leaning against the side of his car. "Crazy dog's so damn determined. Tries to give it, if you'll pardon my French, to other dogs. To cats. Even house shoes." He sighed, in no hurry to leave. "Humped the sister-in-law's foot the other day," he added dreamily. "Never saw a woman get out of a chair so fast."

In the back seat, the terrier strained absently toward his nether parts.

"Shame the surgery was scheduled so soon." The old man grinned. He leaned over conspiratorially. "If he'd kept his balls longer, we might have got her completely out of the house."

Fifteen minutes later, Merion and Annie were still waiting. Annie sniffed snail trails across the sidewalk. Merion was pacing a little from the afternoon's shattered schedule. She got a text: SO SORRY ON MY WAY SOONEST KELLY, or something like that. Merion considered responding DON'T BOTHER, because, castrated-dog theater aside, this was getting ridiculous. But she thought, *What if Kelly is driving now?* She didn't want any part of the accident texting might cause. Merion had already fired one trainer and been stood up by another, and she didn't have anyone else lined up. Before she knew it, she'd paced three circuits of the parking lot, Annie loping easily beside her.

A battered car wheeled into the parking lot. "What a beautiful dog," the girl exclaimed through the open window. She was a rail-thin college-age kid with dark, bobbed hair and a flowered headband, and she burst out of the car in a tumble of dog toys and fast-food cups. She rushed to Merion and Annie with her hand extended. "I'm Kelly. I'm so sorry I'm late. But here we are. Here we are." She made a proud gesture to the battered little building. "Welcome! Welcome! Welcome!"

"Small but efficient," Kelly said of her establishment. She had to throw a hip against the sticking door to open it. She gave Merion and Annie the five-minute grand tour. Just inside, the battered black counter represented reception; a central room had bright pink walls and gray linoleum, and a row of peeling red dinette chairs was the training area. Catching Merion's curious glance at the orange traffic cones, Kelly said they had both a training and a safety function. Clients and their dogs could work on heeling skills around them. Also — she pointed out, lifting up a cone — they covered up the holes where the hairdressing chairs once were. "Hope to get those patched soon, so we can move the cones wherever we want." In the back, storage room to the left, bathroom to the right, and below the door marked EXIT, a play and potty area for the dogs. Kelly was particularly proud of this last. Her ex-fiancé had put up the chain link before they broke up ("Goes to show you something good comes out of everything"). Her brother brought in the pea gravel ("My family's behind me one hundred percent"). What they really needed next was a sign. Kelly looked forward to the sign. She made a round gesture with her hands. She had an idea what she wanted: a professional logo ("for when we franchise") and backlit ("for all the potential customers driving past at night").

When they went back inside, it was late. The fading light was unkind, making the place look even more derelict. Merion sensed a falter in Kelly's good cheer. Merion had been trying to resist, but she glanced at her watch. This was the point where they should be finding out more about each other, and a more experienced person in a well-appointed facility would have flashed a checklist and a schedule book while reeling off a confident, professional spiel. But Kelly's bravado wavered. She leaned forward where she sat, her eyes on Annie, who was lying beside Merion's chair.

She said simply, "I like it that Annie chooses to be next to you."

*What now?* Merion wondered. The balance had shifted between them. Merion said, "Tell me how you came to be a dog trainer." Though she'd read Kelly's credentials on the makeshift website, she wanted to know something more.

Kelly had worked with horses before she worked with dogs. Her family had had an equestrian facility when she was growing up. Riding and training, with stables for their own horses and those that students boarded there. She learned to ride a horse bareback before she learned to ride a bike. It was a good life. Then the barn was struck by lightning and went up, along with the stables, taking everything, including thirteen horses they couldn't get to in time. Kelly reflexively ducked her head and raised her hands like she was going to cup her ears but instead wrapped her palms around the back of her neck. It was more than awful. It was the worst she could imagine, Kelly said, because fire is mean. What it doesn't take with flame, it takes with smoke and water.

The insurance company settled up. Grieving the lost horses, the family moved away without looking back. Kelly would like to get past that. She hadn't been on a horse since she was fourteen. But she loved animals — the fire couldn't take that away — and she was good with them. She became a vet tech. She apprenticed with other trainers before taking on clients herself. This place was the product of insurance money given to her by her parents.

Dusk gave way to night. The pink walls darkened to a tired red. Kelly flicked on a row of fluorescent lights, making the orange traffic cones suddenly glow. *Dept. of Public Safety*, Merion read along the bottom rim of one of them, and she couldn't help wondering how exactly these cones had been acquired.

Nothing about this looked good. She had been well advised. There were plenty of reasons she should walk away. But Merion looked at Kelly with compassion — a youngster with a new business, a lot of ambition, and a good heart. In a recession. Would she, Merion, have had that much courage?

"How soon can we get started?" she asked.

Kelly said: "Matter of fact, we're open right now."

Merion tells me trust is as important as affinity. Trust in the dog. Trust in the trainer. And after her husband died, she was not a very trusting person. But now here she was in a sink-or-swim situation with the would-be assistance dog, and though the two humans had next to nothing in common, Merion could talk easily to this kid stranger. Kelly gave her the more comfortable of two red dinette chairs, the one without the peeling vinyl on the upholstered seat, and she leaned forward as she listened, her chin resting on both hands propped up by elbows on her knees, her eyes on Annie. *Get it all out there,* Merion thought, and she started with the bottom line: "I have PTSD, anxiety, and an obsessive-compulsive disorder that I need to stop before it takes over my life. I worry too much; I watch the clock too much. It's interfering — it's giving me — it's getting me into trouble at work. When I get very anxious, I think something bad's about to happen, and I can't stop pacing. I need Annie to help me control it."

Kelly didn't seem fazed. The dog's basic obedience seemed pretty good. Maybe it would help Merion most to get right to work on the pacing. Break it down into a set of steps to teach the dog. What should they teach Annie to recognize? When did normal walking become a pace? Merion had to think about that. She was sure there was some dividing line between normal behavior and aberrant. Merion realized that the journal her therapist kept suggesting she keep (and that she kept avoiding, for whatever reason) would now be a necessity. It was not enough to know she always felt doom approaching. It was not enough to say "I get anxious, I pace, I'm always worried about time." Merion felt a half-sick thrill at the size of the job before all of them. She had to teach Annie to read her, which meant Merion had to learn that forbidden language too.

Merion paced when she was chasing time. Merion paced when she was anxious about the safety of her school. Two simple concepts, but she says it took her a few weeks to really figure that out. It is always hard to be objective in the thick of things. At home, her pacing was worst when her schedule was at the mercy of others: when she was waiting for a UPS delivery or for her daughters to bring the grandkids. A half a day waiting for a repairman — especially if that repairman's arrival fell outside the hell of that four-hour window — could render her pretty much useless. She chafed and paced when she could have been getting spontaneous things done. At school, where everything was marvelously structured to time (and now she wonders if her choice of profession had something to do with that), she paced when some uneasy something made her worry for her students. Some days were worse than others, and though it is still hard for her to use the term *obsession*, she has to. Merion is much less bound by her condition now, but she says she still thinks of it in the present tense. Even though her work with Annie has kept episodes at bay for almost a year, Merion says she's probably healthier if she stays aware that her condition is out there. It was bad once; it could get bad again.

For some people, she's heard, the pacing makes them feel better. For Merion, it did not. She thinks it might even have made her feel worse, this loss of control, this useless waste of time.

She was eager for Annie's help. Working with Kelly, Merion decided that she would like the dog to interrupt her pacing any time she, Merion, made more than two circuits of the same area. She wasn't eager for the dog to intervene when she returned for something forgotten, but in thinking over her own behavior, Merion recognized that she tended to pace without pauses, a course in any given place that might be repeated thirty times or fifty times or until the repairman or whoever finally arrived.

After at least two rounds of the same space, Merion wanted Annie to block her — to literally stand in her way. When her therapist suggested that Merion might have better success with a redirection, an alternative behavior, Merion decided she would train Annie to block her and offer Merion Annie's leash. Merion's pacing would be the cue for Annie to block. Annie's block would be the message to Merion that she was pacing. Annie's leash would be the redirection to more productive behavior. A walk for Annie. A step into fresh air for them both. At first it seemed odd to redirect walking with more walking, but Merion says it wasn't too difficult to see that a walk with a purpose was very different from the anxious back-and-forth.

This meant teaching Annie to recognize the pace, interrupt the pace, pick up her leash, and bring it to Merion. The four tasks could have seemed huge, but they reminded Merion a bit of teaching preschoolers to square-dance. The little children learned the routine eventually without the calls, learned to recognize certain passages of music, then move a certain way, join hands here, turn there.

Kelly thought Annie would be quick to learn her tasks. She said, "How many dogs do you know who hear the sound of their human's car and run to the door, expecting something good to happen?" Merion didn't know any, but Kelly laughed and said many, many dogs did this. So now it would just be a matter of teaching Annie what to expect and what to do next and demonstrating that something good would happen when she did it. First she would be given a treat for the block, and then a walk. Since Annie was a dog who loved the outdoors, this reward would be right up her alley.

Kelly gave Merion a clicker. Its sound would be a clean, crisp marker for Annie that confirmed for her when she'd done the desired thing: recognized the pace, made the move to block, presented the leash. This was a training tool that they wouldn't have to use forever. Merion clicked the little object and immediately liked it somehow, the cheerful, chirpy sound, like the Halloween noisemakers of her childhood.

Merion paces. Annie blocks. Annie offers leash. Merion and Annie go for a walk. That's the sequence.

"What do I do when it's raining?" Merion mused.

Kelly said, "Take an umbrella."

Merion's therapist said, "Right on."

Six weeks later, they were firm collaborators. Merion would stand in the Great Hall of Traffic Cones, as she called it, and Annie would lie on the floor while Kelly sat in a chair at the far side of the room. Sometimes Merion would pick up some random object, say an empty box, cross the room, and put it on the counter. Annie watched. Merion would move to the window and stand for a long moment, looking out. Annie watched. And then somewhere in the training period, Merion began to pace. A slow cross of the room in front of Annie, as though she had somewhere to go, and then a turn and a slightly faster cross back. Annie would look into her face on the second cross, tail *thump-thump*ing on the floor, but when Merion didn't respond and turned to pace again, Annie would rise. She would tolerate two, sometimes three, passes and then move across her path, her attached leash in her mouth.

As they advanced in training, Merion began stopping and acknowledging Annie absently, then sidestepping the dog to keep pacing. It was a big day the first time they tried this, simulating Merion in so deep a funk she resisted diversion. Annie joined her, bewildered, permitted one more pass, then blocked her again when Merion turned to go back the other way.

Merion looked down at her dog, hands on her hips. Annie sat on her feet and grinned upward, as if daring her to try to pace again. Merion thought the only way the dog could have been more obvious was if she'd set off fireworks. She blocked Merion's pacing when she should have, and when she was ignored, she blocked Merion again. *Intelligent disobedience* is the term handlers use for a dog's awareness of a greater goal that requires a momentary refusal to do something a human partner asks. That moment, that willful intercession of the dog, and the equilibrium for Merion it provided took them six weeks to achieve.

It began with something as simple as giving Annie a cheddar treat after every second turn in a pacing sequence. She was not rewarded for intervening too early. She was clicked, rewarded, and praised every time she got it right. Merion noted in her journal that she, too, seemed to have been classically conditioned. Sometimes when she started pacing, she got the vaguest whiff of Annie's cheddar treat, anticipated the scent before it happened. Merion wonders if now anytime she smells cheddar, she'll stop where she stands.

Puzzle determines a missing person's direction of travel on a testing scenario.

Smokey and Misty arrive one day after their beloved owner died.

It took time for anxious, complicated Smokey to relearn how to play.

Misty, the happy extrovert.

Gentle senior rescue Sam had known human abuse but retained a loving heart.

Mr. Sprits'l can be trusted for play-by-play commentary, whether it's wanted or not.

Starving Jake was
in bad shape when
he arrived.

The shed that may have held abandoned Jake
before his escape.
RIGHT: Though she never had puppies of her
own, Puzzle mothered Jake diligently.

Mizzen: the chocolate Pom with a voice like Winston Churchill.

The Great Cheese Leave-It. Whose leave-it is bombproof, and who is the reluctant saint?

Ready for adventure: Jake and Puzzle like training together.

"Roscoe played ball the way Cal Ripken played ball."
LEFT: Service dog Roscoe ready to take on the
wilderness.

Adolescent Jake takes
easily to his vest and
its responsibilities.

Safe in arms: rescued Ollie with Tricia Helfer.

Within hours of rescue, Jonathan Marshall takes Ollie for a thorough check-up at the vet.

Out of the shelter and on his way to the vet, Ollie enjoys a moment of sunshine.

Resilient Ollie settled into his new life happily, learning the proper hour to bark for breakfast in less than a week.

Nancy and her service dog, rescued golden retriever Lexie.

Out of a shelter and into a happy working life beside children, Caro has a presence that encourages young readers.

Adult Jake attends illness in other dogs and humans.

After her long, troubling illness, it's good to have Puzzle home.

Working Jake Piper demonstrates good manners and focus at a restaurant.

The upside of rescue evaluations: meeting happy, outgoing dogs like this one.

# 14

SUMMER GIVES WAY TO AUTUMN. The animals and I are ready for it, exhausted from months of heat. I sympathize with the brown, drought-weary leaves that drop early, ugly rather than brilliant. And I'm grateful for the series of violent storms that shake us up and wash us clean. Seasonal change can be an unsettling, dangerous time in central Texas — we have the tornadoes to prove it — and when I was a child, storms terrified me. But now, even the worst of our storms are somehow invigorating. The civil-defense sirens wail. We scurry to the inner rooms of my old house, and I sit there, sometimes for more than an hour, amid the sweet, warm breath of dogs and cats who seem to enjoy the huddle, dozing on one another, sometimes draped across my lap. Only Mr. Sprits'l, my single storm-phobic dog, chitters and worries over the thunder. Jake Piper is unconcerned. Puzzle and the rest of the Poms sleep through it. Afterward, the cold air is welcome. It smells like change, and even the senior dogs turn frisky.

Jake Piper's coat begins to come in. He'd been patchy and bare from malnutrition when he came to us, but now several months of good food have put fur on him. And muscle. While I can't figure out how big he's going to be — his legs are long; his feet are small; his head outsize compared to the rest of his body — I can tell Jake's going to be powerful. The chase play with Puzzle is more evenly matched. She is still stronger and corners much better, but when he can manage to stay upright, Jake is gaining speed. Puzzle has to struggle harder to take him down. He came in frail three months ago, but now he is almost the strongest dog in the house.

A dog like Jake could be hell on wheels very soon, but something in his nature prevents that. He reminds me a bit of Gene's Merlin, with his gentleness to the little dogs, and there is something in his loving, connected eye gaze with humans that makes me think of Bob's Haska.

Despite his size and some remaining puppy immaturity, Jake Piper is still what I would call a soft boy. When Puzzle and he are roughhousing, he struggles for dominance, but in every other thing he defers to her. He defers to all the dogs, in fact: young and old; big, small, and smallest. I've often seen rescued dogs come in stunned and subdued for the first couple of weeks and then begin to get a little pushy, laying claim to territory — corners, toys, even humans — with a sort of "I'm here, get used to it" swagger that can become a problem. But not Jake. When he's not racing from one end of the house to the other in play, he pads through rooms wary of the other dogs' spaces, cautious not to encroach. If he does cross some invisible boundary a Pom (Mr. Sprits'l) has set, and the Pom (Mr. Sprits'l) springs forward crabbing at him, Jake's ears droop, his head bows, and his tail wags a little whip-whip-whip of apology. He's a social beast and eager to be loved.

"Awww, Jakey P," I say after this kind of exchange, and he ducks his head and pulls up his front lip in a shy flash of grin that some dogs do, a sheepish "my bad" when he's not at fault at all. (At least, not from my point of view. The muttering Poms, the cats staring poisonously from high spaces, no doubt think differently. They are sick of his galumphing. They wish him gone.)

Jake finds comfort in Puzzle, who mothers him, and he still clearly needs her. At six months old, Jake suddenly becomes afraid of the dark. A summer-born puppy, he came into a world of long days and short nights, and when he first joined the family, his final outing before bedtime was still reasonably light. But now the time has changed and darkness comes earlier. Jake grows dispirited after sunset. Is there a seasonal affective disorder for dogs? I wonder. Is there such a thing for dogs as the night-gloom common in human depression? I put on all the lights in the house, which seems to help him, but the darkness outside is something I can't fix.

Jake is housebroken, bless him, and he seems to want desperately to be good, but I notice that now at night he's unwilling to go outside without a human beside him, not even in the presence of the other dogs. It's an odd sight, that last nighttime constitutional: Puzzle's calm descent down the stairs, the little Poms flashing past her into the gloom of the backyard, and muscular, adolescent Jake running ahead of me out onto the deck, then stopping to look behind him. If I don't follow, and especially if I turn away, he runs back to me — bladder full and face anxious.

I've never had a dog that was afraid of the dark. I've had plenty of dogs go through expected puppy-fear stages, and the training guides say Jake's on time for one now, but this is something different. It doesn't take much to remember Jake's puppyhood five blocks away: tied down out of sunlight in the deep shadows of a shed, abandoned and left to die there. I'm not sure if this new fear is an echo of that hard beginning or something else entirely, but darkness now seems to have a meaning for him I have never seen in another dog.

For a couple of weeks I go out with Jake at night, both of us orbited by the other dogs, and I stay in his sightline while he does his good-dog business. But the little Poms are quicker to get things done, and patience is not their strong suit. When they're ready to go in, they are Ready to Go In. I hear them spinning and nattering away at each other at the door, and I hear them yapping at me. *In! We. Would. In!* If I move away from Jake to let them back in the house, he will follow me, no matter what the state of his bowels. Reinforcing Jake's housebreaking is the first priority. I infuriate the Poms by making them wait. Sometimes they stand peering at Jake over the edge of the deck, as if to say:

*Are . . . you . . . not . . . done . . . YET?*

As time passes, however, Puzzle seems to recognize the problem. She changes her own routine to stay near Jake at night. Puzzle is flawlessly housebroken and fearless in the dark, and where once she would have done her business and then ambled the perimeter of the fence line, inspecting and sniffing for change, now she stations herself within feet of Jake, a mother-that-never-was attending an overlarge puppy-that-isn't-hers. She doesn't hover, but she remains a calm presence nearby, cropping grass or flopping down on the slate pavers to huff over some scent on the stones.

I have no idea why she does this. It's a choice Puzzle makes undirected by me and based on dog impulses alone. Perhaps she sees his anxiety or senses it some other way. Perhaps Jake's fear has a scent that brings out the maternal in her. Whatever the reason, it's working. I notice that with Puzzle nearby, Jake begins to need my presence less. I'm soon able to back away and let the Poms in the house. Within days, I'm able to merely stand at the door and let all of them in and out at will. Jake's fear of the dark diminishes. By the time he turns eight months old, he is comfortable in the deep shadows of the backyard beside Puzzle. They go out together often and spend hours in the dark. Puzzle seems to lead the dance. She moves, and he is her devoted attendant. I look out the window and see the golden and Jake, white as moonlight, provoking night rodents across the stiff grass.

Jake seems to have learned he can trust his new life. He has food, safety, family. Humans leave and come back again, and in the company of other dogs, he is never deserted, never entirely alone. In this happy atmosphere, Jake Piper grows in every direction. Curious, joyful, engaged; his cheerful personality makes him seem larger than he really is. A dog-loving friend says it's hard not to smile at Jake Piper, because he's always smiling at you.

When we head out into the neighborhood, he is delighted with everyone — dog, cat, and human alike. Especially human. Puzzle and Jake Piper are an odd couple on walks. They are not what people expect. Jake is as social as Puzzle is quietly reserved. Jake likes children and especially loves babies, leaning in when comfortable parents allow it, wagging furiously when infants gurgle or laugh. A few passersby are surprised to see the remote but sweet-faced golden retriever walking with the muscular pit bull mix who approaches everyone with the soft-eyed wiggliness of a pup.

Jake's still learning. He is not a perfect dog by any means. He can pop out of his Sit when meeting friendly strangers. He sometimes jumps on visitors to the house. He is, occasionally, conveniently deaf to his commands. But these are issues I could expect with any maturing dog, and I prefer these to the troubling alternatives adolescent Jake might have shown: the separation anxiety, the resource guarding, the mouthiness that can come from a rescued, unsocialized puppy who's known profound neglect.

Jake also possesses a gift that some dogs have, the ability to recognize physical frailty. Two senior, special-needs rescues come into the house, and he welcomes them with a gentle lick of the ears and nothing more. When Jake greets a convalescing friend quietly and lays his head on her knee, holding there, motionless, she is as grateful for his restraint as she is for his attention. She is sound sensitive and light sensitive after a bout of flu followed by a three-day migraine, and she can barely speak above a whisper. As her fingers move lightly across Jake's shoulders, she wonders, as I do, if this is the service dog candidate I've been looking for. "He sure seems willing," she says. "He sure seems to know what to do."

Early, too early, in the morning, and the dogs are going nuts over something at the front of the house. They are in full voice.

"Dogs! *What* is going *on?*" I shout. In the half-light through the windows, I stumble out of bed and promptly walk into a wall. I still often wake thinking I'm in the old Dallas place, and now it takes me a minute to get out of my bedroom and down the corridor to where the dogs are, near the front door.

Mr. Sprits'l spins toward me, very much aggrieved, then turns back to the window to bark some more. The other dogs ignore me in favor of whatever is outside.

It's a chicken. A hen has hopped the low picket fence around my house and is calmly working her way through the foggy yard, ghosting in and out of sight. Though I can't imagine the chicken would have made much noise, clearly it made enough to wake, delight, infuriate the dogs. A chicken! They are following it from window to window across the house, all of them barking except Jake Piper. This surprises me. Jake's got plenty of prey drive, but he watches the chicken with only mild interest, wagging idly, as though he's a been-there-seen-that country boy to the others' citified astonishment.

When did Jake Piper ever know a chicken?

Through the glass, I can see a lady roll up in a truck. She is driving with one hand, talking on a cell phone, cruising slowly, her eye on the hen. She claps the phone shut, stops the truck, and gingerly gets out. She's wearing sweatpants, house shoes, and a *Dark Side of the Moon* T-shirt. Her hair is every which way, and even when she crouches down and creeps along the fence line in the fog, I can see the spikes of it sticking up. It's like something from *Mutual of Omaha's Wild Kingdom:* No sudden moves. Don't spook the chicken.

My dogs can see the woman too. Bonus! Puzzle bounds easily over the moving hustle of shorter, squabbling Pomeranians. As far as I know, none of the Poms has ever seen a chicken; all of them are ready to tell the lady off. It's a cartoon moment. The little dogs desperately pile on top of one another for an advantage at the window, and the hen cocks an eye, then ignores them completely and pecks her way across the lawn. Outside my front gate, the woman appears less confident. She looks left and right and then at my front door. She seems unversed in catching chickens and a little taken aback by the roar of dogs in the house. Mr. Sprits'l enjoys jumping up to give her the stink eye through a window before dropping to the floor again. *Ha!* he barks triumphantly at the top of the jump. *Ha!*

Meanwhile, the chicken meanders through the front garden, finds something tasty, and snaps it up. The woman can't bring herself to trespass in my yard. She waves her arms and tries to herd the chicken from the sidewalk. The hen bounces onto the opposite fence, pauses a moment, then scurries across the road and disappears into the fog. The woman trots behind her with down-stretched pleading hands.

It's all so ridiculous that I can't help but flop down on the couch and laugh. It's the first time I've laughed hard since Misty's death, and the release is welcome. The dogs don't quite know what to make of this. They circle me curiously, with tentative wags. Puzzle gives my face a sniff-over. Jake puts his paws on my forearm and looks into my eyes. We live, I tell them, where chickens actually do cross the road. Dallas never gave us this kind of action.

I lead the dogs back to the kitchen to make coffee for me and breakfast for them, feeling the fine bones of this old house beneath my bare feet, and I think the move has been good for all of us. The pace is slower here, the people friendly. The traffic noise and the increasing presence of neighborhood crime, so long a backdrop to my life in Dallas, are pretty much nonexistent here. This was a move I'd waited twenty years for, and by luck or fate or good karma, it's just the kind of place, I think, to partner working dogs of every kind. I look down at Jake and recognize that he's a part of that good fortune — the unexpected, unlikely stray I had tried to resist whose resilience and mysteries of heart have something to teach me about my own.

After breakfast, the Poms always want a doze, but the big dogs are ready for adventure. I leash up Puzzle and Jake Piper. It's still early, but we'll head out for some socialization work, some obedience work, and the sweet, simple opportunity for them to put paws to damp grass. When he sees the leashes, Jake levitates once and squeals his delight, but outside he sits lightly beside Puzzle on the first command while I lock the door. Puzzle holds her Sit patiently, lifting her nose a little. Jake's Sit is more obviously strained. There is so much good to explore — now! He looks up at me with a poignant expression, wrestling obedience against the pleasure of a mad dash into the chicken-scented fog.

# 15

FIND *ONE*, PAULA HAD SAID, and Jake Piper may be that one, but now I've been asked to find another. In addition to a potential assistance dog, I've been looking for a dog that might be a candidate for a therapy partner in schools and hospitals or perhaps a wellness companion — a sort of ESA on loan, a comfort-providing pet for an ill person on day visits. I've had requests for help finding both. The health-care professionals who've asked are particularly interested in visits from "lap dogs that are happy to be there." At forty pounds, Jake is no lap dog.

While Fo'c'sle Jack has actually passed therapy testing (greased with plenty of treats) and has worked with children in this way before, Jack is not really happy sitting on human laps. Fragile, anxious, or simply too high-strung, the other Poms are not strong candidates either. Misty alone would have adored it. She would have shone at this. As I'm looking to rescue a dog with the affectionate, easy companionship needed here, a dog completely open to frail strangers, I feel the loss of that sweet girl more keenly.

A handful of rescue friends know I'm searching, and, Petfinder addicts the way I was once a crime-map addict, several of them on the same day point me to a Petfinder ad for a little female Pomeranian. Another hardship case, she was dumped beside a highway not too far out of town. Guesstimates put her at about eight years old. A chocolate-brown Pom with white paws.

*Sweet, friendly, housebroken, small. Cuddly. Good with other dogs. Good with children. Has not been tried with cats.*

I decide to pay a visit to Miss Chocolate Pom. I meet her through a rescue group in a small room puppy-gated off from a horde of other little dogs. Hundreds of dogs, it seems like, though it is probably only about twenty, a riot of fuzzy dogs of all kinds who paw at the barrier and bark for the attention they are not getting from me. Chocolate Pom seems oblivious to them, so oblivious that at first I think she might be deaf, but then I see her ears flick at the faraway opening of a door, and I realize that she's either uninterested in the other dogs or, stunned by the cacophony, a bit numb to them.

She is a funny little creature, the color of a Milk Dud, with a white muzzle that looks like she dunked it in milk. Her once thick coat has been shaved off — and with good reason. It was matted to the point of ulcerating her skin when she was found. Many Pomeranian owners recognize this lion cut with approval or dread, finding the fluffy head, fluffy tail, and bare body a kindly action for a thick-coated dog in hot weather or a terrible imposition on a dog that could be, should be, beautiful. But in this case, the cut was humane, and the groomer who volunteered her services attempted to make amends for the drastic shave-down by painting the dog's toenails pink and putting a matching bow in her hair. I am not sure it helped. With the haircut and her chocolate eyes and pale muzzle, she looks most immediately like a marmoset in drag.

"Hey, monkey-face," I blurt out, because seriously, that's the first thing that comes to mind, but I say it kindly (I hope). The adoption supervisor looks on. I hold out my hand. The little brown dog toddles toward me with a stiff, awkward gait. Bad knees there, caused by luxating patellae, but she is a happy soul. She has more of a prance than a hobble. What little puff of tail she has left is wiggling wildly.

I pick her up, cradle her in my arms, and rub her belly. The tail keeps wagging. I put her down. She skitters around my ankles.

*Hoo. Hoor. Hoo. Aahh-hooooor,* she chatters. It's a throaty, growly little rumble. I have never heard a Pomeranian make such sounds. (I can only imagine how it will set my crew off.) This kid is wiggly and playful. When I extend a forefinger, she pats it with a paw. When I make a light grab at the paw, she removes it from my fingers and play-bows and chitters some more. *Hoor! Hoor! Rrrrr!* She slaps my forefinger again. Grab-toes is the game Jack and Misty loved, a game many, many Pomeranians enjoy, an engagement with humans that doesn't take much strength to play, and the little dog comes to it naturally.

As far as the adoption supervisor is aware, this Pom hasn't had any obedience training. It's hard here, she says, with so many dogs and so little space and not enough time to give every dog the kind of advantage training would bring. But she's sweet, the supervisor tells me, and she isn't ever cross with the others and has never snapped at the adults or the preteens who sometimes volunteer here. I move to where the supervisor stands in order to hear better over the roar of the other dogs, and the little Pom follows immediately.

The adoption supervisor, who looks tired, says the Pom knows this room. She has been here awhile and has been passed over before. "She doesn't want you to leave her," the lady says, a statement that may or may not be genuine but is certainly strategic.

We are left alone for a time so that I can make a considered decision, and the cheerful little Pom never leaves my side. There's a magazine on a table. I drop it on the floor with a soft slap. The dog walks over to inspect it, then comes back to me with that odd hooting. I crumple up a subscription form that has fallen out and toss it. Miss Chocolate Pom tilts her head a moment, like that's a new one on her, then skitters to the ball of paper, gives it a sniff, and tentatively picks it up. I hold out my hands and say, "Bring it back," and though I know this is all unfamiliar to her, I can see she's torn, taking a few steps toward me but finding it impossible to both carry the paper and hoot, which she clearly wants to do. She drops the wad of paper midway and dances back, flopping down at my feet and rolling her belly upward. It is the ultimate sell job, and I'm sold, knowing that even if for some reason this gifted little candidate can't become a therapy dog, it is time for her to have a home.

Introductions at our house are an old routine. We've had dogs in transport visit for as little as two hours. We've had foster dogs stay as long as five years. So the little chocolate Pom, whom I've named Mizzen, comes in the way they all do, carried through the house and put in a room to get a sense of settle and then introduced to the others, one by one. If the solo introductions are no big deal on all sides (and sometimes, through some strange alchemy of personalities, they aren't a big deal, with the old-timers barely blinking at the new guy), they are all in Sits for treats in the kitchen in just a few hours. I've had some magical moments with some of the transport dogs, little old blind or arthritic creatures, dazed with uncertainty and change, who in some shadowy past learned Sit. The word somehow grounded and reassured them, and they plopped down with my group in the kitchen as though saying, *We know the drill, thank you, and bring on the cookies.*

Even the new dogs with no obedience training are quick to learn Sit in the company of the other dogs. They learn how goodness pays in treats. Mizzen wants to be one of those dogs. She is completely food motivated, and she understands that the kitchen is where it all happens. She figures that out straightaway. But Sit doesn't come quickly. Is it the intellect or the bad knees? She bounces up with the other dogs when I approach the treat drawer, but when I say "Sit" (and dog backsides, large and small, hit the ground), she makes no immediate connection to herself. They sit. They get treats. She capers, bounces into the other dogs, spins a little, play-bows, says, *Hooor! Hoor! Rrr-hoo!*

No treat.

I tap with a forefinger right above her tail. Hold the treat over her head in such a way that most dogs would have to sit to look up at it.

No sit.

No treat.

*Hoor! Hrrr! Hoor!* she mutters. *Aaahhhh hoorrrrrr!*

("She sounds like Winston Churchill," a history-buff friend comments later. "'The farrrrrther backward you can look, the farrrrrther forrrrrward you are likely to see.'")

"Sit!" I try again. And again. And again. And then one day, I lightly touch her hips while holding the treat over her head, and she magically sits, beaming up at me, then snatches the treat with a *hoor!* of happy desperation.

She sits on the first command ever after.

I'm especially interested in Jake Piper's reaction to the new little Pom. Dog-friendly he has always seemed to be at home and abroad, but he's the puppy of the family, the one who has starved, the one who has had Sit in his repertoire the least amount of time. He has been center stage, too, all this time. Though Jake has never shown any signs of aggression, small dogs can be provocative for big ones — new small dogs maybe even more so. Whether he frames the little dogs as pack siblings or small, chaseable critters, I can't really know, so Jake is the boy whose response I'm watching now.

But when it's treat time and Mizzen spins and pogos off Jake Piper, he barely seems to notice her. The other Poms grumble over her heedless skittering, but Jake seems literally and figuratively above all that, treating her with polite deference. He is frankly uninterested. For whatever reason, Mizzen comes to adore him. While the other Poms chide them both and Puzzle watches with what seems to be gentle amusement, Mizzen is like some love-drugged character out of Shakespeare. She is clumsy with infatuation for Jake Piper, hooting and tap-dancing after him as he moves through the house. Where I might find her underfoot presence annoying, Jake Piper is all equanimity. He moves left. She moves left. He flops down in the hallway. She flops down in the hallway, gazing at him with limpid marmoset eyes. Sometimes when they are moving quickly together, he suddenly stops, and she can't brake as easily, so she keeps going on the polished hardwood floors. More than once she's plowed into him or slid under his belly to the other side.

*Hoor!* she rumbles good-naturedly, careening back.

Every once in a while, Jake Piper gives her a kindly lick.

"So odd," says one friend about them both.

"Her wiring is off," says another.

"And he's really good for a pit bull mix."

I believe Jake's patience is really good for *any* dog, and I say so. Mizzen is a sweet curiosity. Jake is a daily surprise.

I am most impressed one evening when the whole dog-and-human lot of us are piled in the living room watching television, and Mizzen is doing everything she can to win Jake's attention. She nuzzles his paws and burrows into his chest. She lies spine to spine, then comes around and snuggles up against his belly. He is patient. At one point I hear Mizzen hooting happily and a strange squeak out of Jake Piper, and I look up to see Mizzen pressing her head against Jake's mouth. Pressing, pressing, pressing her head against his mouth. This is new. She is insistent, a little battering ram storming the castle, and after a few bewildered moments, Jake opens his mouth and Mizzen crams her fuzzy head inside it. I can hear a pleased rumble in her throat.

*Yaagh,* he gurgles after a moment.

She withdraws her head and then does it again.

What is she doing? I have no idea. But kindly, bewildered Jake acquiesces to her say-ah weirdness. Ears swung back, his eyes wide, he holds still while the little Pom heads in to rummage through his mouth, mumbling and crooning. But he looks at me dubiously, this good dog, as if to say, *I sure could use some help here, if you have any ideas.*

Use Your Words is a dog trainer's mantra, but Mizzen doesn't have many she understands, and this is outside my experience. What do you say to a dog with this kind of obsession?

I gently extract her. Jake shuts his mouth. "Mizzen," I say, "stop feeding your head to your brother."

We have no idea I'll be saying this many, many times.

# 16

JAKE PIPER'S GOT AN audience here in the center of town. The local restaurants are doing a brisk Saturday-breakfast trade. The shops have begun to open, their plate-glass windows reflecting an uneasy sky. It is a day that could pucker up and rain without much warning, and those of us out here are taking advantage of every sunny second.

Jake's on a dropped leash at the moment, holding his Sit on the sidewalk and looking wistfully from me to a young family seated on a bench a few yards away. He would very much like to get up. He would like to break this Sit and leave this Stay. They are all kinds of tempting, this family — a pretty young mother and her husband sharing an ice cream cone, the mother carefully cradling her sleeping baby. They both watch Jake train with a liking so obvious that even if he can't see them in periphery, he can probably feel or smell their approval. Dog people. Dog people with a blanketed, invisible baby, a dog pupa. Jake would very much like to woo them.

We've been meeting a lot of strangers lately, but as far as I know, Jake has never met a baby this small, and there is something about the bundle of infant that intrigues him.

"You can see the wheels turning," says a passerby with a laugh, because there's nothing relaxed about this Sit. Jake's bottom is down, but his thigh muscles are tight, and the tail *swoosh-swoosh*es with any kind of acknowledgment. He likes human voices. He likes human glances. The upside of Jake's great eye contact is that he recognizes my warning human eye during a training session. The downside of Jake's great eye contact is that he immediately picks up on dog-loving gazes from strangers, and even silent supporters of his training can wind him up with mere friendly looks.

That passerby is right. If Jake were a muscle car, you'd be able to hear his engine revving.

"Jakey, *come*," I say to him, and he springs to his paws a little wildly, one ear up, one ear sideways, all goofy, pinwheeling paws and tail. But I see his footsteps waver; he is charged with conflicting impulses, going to me or veering straight for this splendid family full of goodwill and streaked with the ice cream that has run down their hands in the humidity.

He comes to me, but only just, and when he gets here, I can feel the regret. He sits again, shoots a yearning glance over his shoulder at the family, then looks back at me.

"I give you an 8.5 on the dismount," I tell him, slipping him a training treat. "But good boy. Good boy!"

The family wants to meet him. They call out across the sidewalk to us, but the current tension on the lead tells me Jake Piper is on a hair trigger. He's a friendly boy, but he's still learning. He's strong and standing up, and they are seated, wrangling ice cream and an infant. I tell them we will head down the sidewalk to work on Heel, then make a circle and return. If they've got time to wait, we'll give them time to finish the cone. (And reconsider. And double-check their purchase on the baby. I don't say this, but I think it hard enough.) We'll give Jake time to work off some of the excitement too. Jake Piper's got to earn the opportunity to meet them.

Jake takes the command away from them and accepts the Heel with better grace than Puzzle ever did at his age. I expected so much fight on a leash from this boy, and while he can lose his head and pull now and again, he seems to see walking on-lead as a joint adventure. Puzzle as a youngster took to the lead grudgingly, accepting our attachment more because she had to than because she wanted to. For Jake, everything is better because it's shared.

We return to the family, approaching them from behind. Even though Jake has done a fine walk away from them, I'm a little nervous about this meeting, knowing Jake's size and his strength and his bull-in-a-china-shop love for all humans. There's a fine line when you're learning to trust a young dog. You want to be aware, very much aware, but you don't want to transmit your own tension down the lead. The dog feels it and can feed off your energy. When we circle the bench, we can see both mother and father bent over the baby on the young woman's lap. They are talking low, smiling into the blanket at a face we cannot see. But a tiny wiggle shivers the fold of flannel, and at the sight of that movement, Jake's ears perk and his tail whip-whip-whips.

"Here we are," I say to the parents quietly, not wanting to rev up Jake any more.

They look over and smile. Jake's tail speeds up.

Jake approaches, pulling the lead tight, and just at the moment I'm about to tell him, "Easy, easy," he sits. He puts his head on the knee of the young woman.

*Chuff,* he huffs from his nose, scenting deeply.

"Well, hello," she says and smiles. Her arms are full of baby, but her husband reaches over to stroke Jake's head.

"Crazy ears," he says.

"They are."

"Good dog, though."

"Thank you. His name is Jake Piper."

Jake has not moved from his position at the woman's knee. I can see his eyes working, looking from person to person across the conversation. He is listening to the adults and watching them, too, but his nose is extended toward the baby. The nostrils quiver and flare. He is curious. He would like to sniff more closely, would like to touch, but something in Jake Piper holds back, as though he needs permission. Who is this dog? I was ready to correct a youngster that might be too overjoyed by new people, but I was unprepared for Jake's response here. It's as though he has absorbed this family's moment, and, with some need to be drawn into their circle, he has quieted himself to meet them where they are.

The young mother says to Jake, "This is Emma," and to me, "Today is a big day for us. Our first day out. Emma was born with pneumonia two months ago. We almost lost her. The doctors finally said we could leave the house with her last week, but we waited . . . and so . . . this is our first day out." Her voice is tired, joyful, relieved — all those things in four words: *our first day out.*

We are proud to share it with them, and I say so. I wish I'd taught Jake to offer a high-five.

Jake lifts his head from the woman's knee and leans it into the young father's hand, groaning slightly at the ear rub. He is as calm now as he was wired earlier. The visit with strangers, the obvious approval, the ear rub, is enough.

Baby Emma's foot wriggles free of the blanket, and from where he sits, Jake lifts his head and raises his eyebrows. Whoa! The baby is strange and wondrous. His tail wags small, tentative, as though amazed such tiny human things could ever be.

Cheese or goodness. Cheese. Or goodness. Jake Piper and Puzzle are in Down/Stay commands on the back deck, and I have dropped a chunk of cheese and told them both, "Leave It." The cheese is a chunk of smoked Gouda, appealing by scent even to me, and it's in easy snatch range for both of them. They wouldn't even have to break their Down/Stays to do it. Puzzle doesn't budge, grinning up at me with happy affability. She's been Leaving It for years, and she's brilliant at it now, though I'm aware that this has always been easy for her — unlike Jake Piper, she's never had to scavenge for food. She likes cheese, *but hey,* she seems to say with a shrug, *whatever.*

Jake Piper finds this much more difficult. He is as food motivated as they come. Unlike Mr. Sprits'l or the golden, who tend to sniff offered food thoughtfully before taking it, a critical assessment of its appeal, Jake eats first and asks questions later. And now, even though he's been fed today and never goes without a meal, this cheese torments him from its position fifteen inches away. Cheese. Or goodness. He extends his forefeet and sinks into his Down, looking up at me with a darkly pained expression, gripping the deck with his paws as though he's having to hold himself back. Together, the dogs obey the Leave It for ninety seconds, and then I call them away for a toss of the ball. After they scramble off for it, I pick up the cheese and throw it away. Bystanders have asked why I don't give it to them after the training moment, but it's a mixed message to reward the dogs with the very thing they've been so good to leave. Fortunately, Jake loves a ball toss almost as much as he loves cheese, and Puzzle loves motion much more than she loves food of any kind, so this reward is enough.

Leave It compliance is a must for a search dog, a service dog, a therapy dog, and, in my opinion, for a pet dog too. Not always about food, and not just about good manners, Leave It on a walk can save a dog from dangerous contact with tempting hazards — choke items, infected scat from other animals, rodent bait, dropped medication. Leave It is an important command. For Jake Piper, it's a tough one. In the earliest days, he would snatch the food and run away to eat it. Then, fond of people and fonder of their affection, he would snatch the food and Sit or Down immediately afterward, with a deep pant of a smile — a total masking maneuver, as though the subsequent obedience made up for the earlier indiscretion or, even better, persuaded me that the food snatch had never happened at all. Jake makes me laugh with the workings of his dog mind, but I'm not fooled, and we continue to train on the sweet torture that is Leave It, sometimes with Puzzle and sometimes not.

I find myself sniffing food from the refrigerator and trying to assess its appeal the way a dog does. If dropped food items were ranked, what would be the least tempting to a dog? The most? Should we work our way up from toast to prime rib?

That's the plan. I started with white-bread toast crusts, then whole-wheat toast crusts, then pita and naan and tortilla chips and pappadam from restaurants. I dropped crackers and flung bread sticks. When Jake successfully turned away from whole-wheat toast, which seemed to be his favorite, I made things more difficult by adding butter. One very hard day I dropped half of an Olive Garden bread stick in the side yard we would later pass on our walk. Jake picked up the scent from the ten-yard line, and by the time we got close I could hear him snuffling and feel the lead quiver.

I know humans who might look both ways and then go for a dropped Olive Garden bread stick.

"Leave It," I said, as he caught sight of it. Jake lunged and then thought better of it. Without a check from me, he let it lay. But he flopped down into a Sit and let out a wail of protest, a frustrated Wookiee cry unlike anything I'd ever heard from him. I let him howl it out, and then he stood up, shook, and glanced at me as though he was glad he'd had that good cry. He was ready to go. Off we went. Jake didn't look back.

On the return trip, though the bread stick was still there, he studiously looked away from it and the huddle of pigeons that rose up from the pecking of it.

"That's so mean," says a friend, who admits she doesn't really get the whole working-dog thing. She likes the idea, but she hates that a dog can't score a dropped cookie. I explain that I never taunt the dogs by offering them food and then denying it to them, but I do accidentally drop things from plates, and I pre-drop things privately that we later pass on our walks. I try to explain the safety angle too, but I'm not sure my friend believes me. Would a dog really swallow a dropped pill simply out of reflex? Sadly, many dogs would.

From plain bread to buttered bread to child's food, like cupcakes and cookies, I try to drop what we might find left on a sidewalk or tossed from car windows. I don't have to plant too much: on some of our walks, we pass the greasy remains of paper french fry pockets and slick, meaty, cheesy burrito wrappers. Jake Piper extends his nose toward all of them, then retracts it on the Leave It, once even ignoring a wad of hamburger and a splay of exploded milk shake at a bus stop. Good work. We're getting there.

Which is not to say Jake Piper doesn't explore other food disobedience. Counter-surfing, the great — and dangerous — problem many dog owners have with tall dogs, is a temptation for Jake from the moment he reaches adult height. Puzzle has never counter-surfed, and the Poms, who might like to try it, can't, since they're not tall enough. The counter-surfing dogs of my friends have eaten whole loaves of wrapped bread, steaks left out to thaw, entire pans of brownies. One friend admits that the first stolen sandwich was cute, but when her dog managed to get the Thanksgiving turkey on the floor before it ever reached the family table, it wasn't funny at all. Dangerously, my friends' dogs have also bitten into and exploded plastic bottles of cough syrup and lapped up sugar substitutes, poisoning themselves. There are dogs that have counter-surfed gas stoves and turned on burners.

There is certainly evidence of this behavior in my century-old house. Beneath the lovely refinishing, the wood kitchen counters are scored with hundreds of claw marks along several lengths of counter edge, marking where one family or another left food out and one dog or another went for it. I recall the day when I briefly walked away from the counter in the middle of preparing the dogs' food bowls, and Jake Piper, to his delight, found he could raise up on his hind legs and put his paws on the counter and get to almost every single one of the bowls!

*"Don't you! Don't . . . you . . . even . . ."* — *think about it,* I was going to say when I walked back in and found him, but Jake, startled by my vehemence, dropped his paws and his ears and slunk beneath the kitchen table, putting himself in an apologetic Down/Stay and watching me with a penitent expression. The guiltless but interested little dogs also scrambled. Jake Piper never counter-surfed again. And somehow, that self-discipline has translated even to the Poms. I can leave my own food on low tables, walk out of the room, and come back to find a ring of dogs at a little distance, staring as though they could will cheese fries off the plate.

It's clearly a bigger moment for me than it is for Jake the first time I put on his official In Training vest. He aced the service evaluation; he's learning good manners. When I slip the vest over his head and over his collar, he politely holds his Sit but looks a little thoughtful, wayward ears swinging forward and back independently, as though in dialogue. He turns his head to sniff over his shoulder at it. It's a curiosity, this dark green vest; stiff and a little formal, it fits something like his seat-belt safety harness and probably still smells a lot like puppy Puzzle, who wore it until she certified for search. New vest patches reflect Jake's different job description. WORKING — PLEASE DO NOT PET replaces the ASK TO PET ME, I'M FRIENDLY patch that Puzzle once wore. Jake accepts the vest easily, but it brings a new lesson that will be harder to understand. In the vest he should neither receive petting nor ask for it. Some dogs handle the transition easily. For others, it's a testing point. Extremely friendly dogs that cannot override their impulses to meet everyone and greet everyone at all times aren't appropriate for service. Jake could be one of those dogs. He is smart and eager to please, but I just don't know.

At the heart of any kind of therapeutic service animal is a dog prepared for public good manners. The strong Sit, the willing Down, the self-disciplined Stay, the walk on a leash that doesn't put owners or others in jeopardy, the calm and friendly meeting of other dogs and other people, the leaving of food where it lies — it all makes as much sense for the pet dog as for the assistance or therapy dog, and even if Jake Piper ultimately cannot work service, he is becoming the kind of dog that is a joy to take places. He'll take the Canine Good Citizen test soon.

From the CGC, we'll work toward the Service Dog Public Access Test, which evaluates dogs' obedience and etiquette in interior spaces like stores, restaurants, and airports. Though Jake will never enter such spaces as a fully trained assistance dog unless he is actually serving a disabled human partner, as a service dog in training, he will work with me toward a future that could take many forms. Yes, he may remain with me, but there's a chance that he might be so good a fit for someone who needs him that he will leave us. Jake has no abandonment issues, no separation anxieties. As he has matured and I've gotten to know him, I've come to recognize he is the kind of dog who could leave and make a transition to a new human partner happily.

But could I make that transition as easily? Years ago, I would never have imagined being able to do this — saving a dog, loving a dog, training a dog, and letting him go. I never understood how service dog puppy raisers could do it. I always wondered if puppy raisers somehow kept the dogs at arm's length emotionally, if it was somehow a matter of loving less. No. I realize now that, in a way, it's a matter of loving more. Love built on generosity — that openheartedness is new ground. Jake has it. I'm learning it. Though I know what it is to love a search dog at the same time I follow her into the dark, this love with a view to parting is a stretch.

# 17

WHEN THE YOUNG WOMAN talks to me, her voice wobbles a little, sometimes uncertain, sometimes edged with what sounds like derision, her tone wavering between *screw this* and *whatever*. Kristin is nineteen. Slim like a dancer, with beautiful posture, she has dyed her long hair very black and has pierced a lot of real estate: nose, ears, lip — tongue too, if what I'm hearing when she talks is the sound a stud makes. She is a pretty girl in armor.

Kristin began having "episodes" three years ago. She makes air quotes with her fingers and looks away when she says the word. Episodes, like she's a goddamn TV show where everything will come out right in the end. She leans forward and puts her hands on the table, fingers splayed, and I think, *Discussion over, this girl's outta here, about to push up and away after delivering a parting shot*, but she just sits there, studying her black manicure, and says she was this person once, and now she's someone else. It sounds like the opening of a story she has told before to great effect. What follows sounds ungrammatically defiant: This is her, and this is her owning her, and she doesn't know whether to be proud or ashamed.

Six blocks, she says, describing the first time in her "illness" (air quotes hers) she ever got lost. She just had to make it six blocks, from the convenience store to her house, a convenience store she knew well, the route back to her home a path she'd known all her life. It was late afternoon, almost evening, the kind of light where the pink sky begins to gray at the edges, and she was walking home, and all of a sudden she had no idea where she was. One step fine, the next uncertain, and she looked up and was totally lost. And worse than lost, lost and panicked, unable to take a deep breath and figure things out. Losing your way in your own neighborhood is like being a fish and forgetting how to swim. And that was how it felt to her — not just disorientation, but drowning, as though the entire world were liquid and insubstantial and she was going down.

She did go down too. Smacked her forehead on the sidewalk and lay there in a puddle of diet soda she'd gone to the store to buy, and in confusion she was certain that everything was melting like ice cream and she was melting too, down and down, into the sidewalk and beyond. She tilts her head back now and lifts her hands to make a paddling motion. She says that's how it felt, and she wonders if she looked like a stranded bug, and she wonders how long she lay there. What's curious is that in this friendly town, where everyone seems to know everyone, no one came to help. But maybe no one saw her. She can't imagine people saw her and didn't help. It's not what we do, she says. Even the people who aren't particularly nice would have at least called 911. No one wants a teenager gasping and slobbering on his front walk.

It was dark when she was able to right herself, and when she did, she knew she wasn't far from home. Four houses to the corner. Turn right. Three blocks more, and there would be her house. She was so relieved to know where she was, she almost peed herself. (Okay, she says. Correction: She didn't almost pee herself. That is one problem she doesn't have.) Anyway, she stood on the sidewalk, then gathered up her purse and wobbled home, passing neighbors taking in groceries, kids screaming through their games in the dark. The lost feeling was over, but she still felt vacant, abstracted, the way you do after someone you love very much has died and you have to move through a world that has no idea something terrible has happened. Putting one heavy foot in front of another, Kristin remembers seeing the light of the back porch shining across the wet lawn of her backyard and feeling so relieved when she walked through the gate that she bowed her head to the grass and just smelled the green earth.

Quite a surprise for her dad when he walked out and saw her smelling the ground. The next day they had a Serious Talk about Drugs, but no way was she going to tell her dad what had happened the night before. She nodded and promised to just say no. (And she meant it, she says. Drugs. Seriously? With a head like mine already packed for somewhere else?)

It all became a big deal when she got disoriented in her own high school and fainted in a hallway. Except her eyes were open — can you faint with your eyes open? Well, I did, she says. She remembers looking up at fluorescent lights and ceiling tiles. At first, everyone thought she was just being a drama queen. But when it happened a second and third time, school counselors encouraged her parents to take her for a medical evaluation. The counselors said that if Kristin was putting this on, the performance likely wouldn't survive a couple of cold rounds on an examining table. If there was a medical problem, it could be diagnosed and treated. And if medical tests didn't find anything, a therapist might help.

Her folks love her, she says, and they had no reason not to trust her, but honestly, this was all coming up as a fake to them. She is still angry about that. She's also scared shitless about the dizzy spells that sometimes take her to the ground. Fear seems to make them happen more often and worse. She props one foot on her opposite knee. She picks at the rubber sole of her Converse Chucks, covered in little hearts. A whole lot of doctors, she says. She can't remember them all. A blur of people over her head, not talking straight to her, like she was a thing instead of a human. "'Thing,'" she says, making air quotes again, and looks away.

Medical tests were inconclusive. There was no concrete explanation, but she was prescribed some pills. The pills left her apathetic, and she sank into a dangerous depression. No one had any idea what was happening. She says they still don't. When she fell to the asphalt while crossing a city street and was unable to speak to her father upon waking, the teenager was hospitalized. More tests. More therapists. Kristin was finally diagnosed with acute panic attacks, but medical and mental health professionals couldn't pinpoint the cause and tried a number of treatments that didn't succeed. She participated in talk circles, drew pictures, and role-played with dolls before a series of strangers. She missed the last three months of tenth grade and was told she would probably have to repeat the year.

Kristin describes her episodes in Harry Potter terms and wonders if author Rowling modeled the Dementors on a condition she herself knew all too well — the sudden sink of depression; the unexplainable, overwhelming grief; and the breathless sensation of falling. Kristin can't be sure if she falls before the sensation or after it, but she's planted her face on concrete a few times. She wishes the attacks gave her more warning than they do. She used to be a confident person, but now she no longer trusts herself to do anything alone. She's seen padded helmets for epileptics. She thinks she needs but would never wear one.

When she finally went back to school, Kristin went back skinny. She went back different. Withdrawn and cautious. She threw up her Pop-Tart first thing on the first day, so the word went around she was anorexic. Her friends acted uncertain, and her teachers were suddenly big and cheerful. Aware of their falseness, Kristin felt more isolated. She started dreading school. She began to feign illness in order to skip out entirely. Therapists recommended a school/camp combo for teenagers with similar conditions, but Kristin was walked out and talked out and resistant to the whole idea. Her parents struggled with the decision to send her away against her will. Her mother was afraid that away from home, she would sink too far down to survive.

Even though Kristin's episodes don't happen every day, or even every week, they have totally altered her. She remembers what it was like not to wonder if she'd make it down a hallway without falling; not to wonder if, when she set out for home, or school, or work, she'd actually get there.

She searches for a better way to explain.

She says the first time she saw *Jaws*, she was just a kid. She left her friend's house, where they'd played the movie on a big-ass home theater, and as she walked down the street, shaken, she kept imagining a shark roaring upward out of the asphalt. "That's it! That's it!" she says, stabbing at the table with an index finger. "Now *that's* what it feels like to suddenly be lost. Just when you think you are fine, it rushes up out of nowhere, takes you down and down, and swallows you whole." She sits back, satisfied to have nailed the comparison. Between the Dementors and Bruce the Shark, her episodes make one helluva day.

When a service dog was suggested as a possible support therapy for her condition, the first name she thought of for the dog was Placebo. Fatigue, frustration, and anger made her sarcastic. She told them she wasn't stupid enough to be distracted with a pet. She has seen guide dogs for the blind, and, on television, a documentary piece about seizure-response dogs, but she could not imagine how a dog could do anything useful for her.

Enter Juice Box, who came with another name but quickly earned a nickname from Kristin. Juice Box had previously worked for an elderly man who had balance problems and trouble carrying things, and the man had died. The dog was still young; he was friendly, obedience and service trained, and, as one counselor put it, more than a little lost without something to do. That persuasion seemed transparent. Her parents were more open to the idea of a service dog than she was, even though the family had never had a dog and she had never asked for one. No one would drop it, and Kristin finally agreed just to shut them up.

She had thought the coming dog would be a German shepherd, like the guide dogs she had seen, but when Juice Box arrived, he was a surprise all around. He's a super-confident, big I-don't-know-what, she says. Boxer in there somewhere, maybe Lab. He is brown and weighs almost as much as she does. He has strange rough fur, a huge nose, and then kind of wide but delicate paws. Cute though. This big wide, grinning face. He is big enough to steady her if she gets lightheaded, and he came already able to lead her someplace to sit down if things got bad. The plan was they would teach Juice Box to help her stay safe when she got disoriented, lead her home, maybe, and if he could do those things, it might go a long way to getting Kristin back to the life she had lost.

Juice Box is interesting because he's smart, but he's funny too, and she had not expected that. This dog is a total ham. He is crazy for tricks. He can carry things without dropping them — a cookie on a plate! A TV remote, without changing the channel! It was awesome the first time he did that. He will put his paws up on her shoulders and dance. She says The Juice is totally jacked. The slang goes over my head, but I hear the affection in Kristin's voice.

"What does your dog do most for you?" I ask. She says, "Shit-I-don't-know," straightaway, and then she pauses, thinks about it. She says that even in the small town she's known all her life, she became afraid to go anywhere. She can get so scared about another lost event that she doesn't trust herself to get from A to B. But Kristin trusts Juice Box. They started small: From the yard to the front door. From the street to the front door. From one house away to home, and then from two.

Kristin has passed her neighbors' houses with such attention that she knows what hours their sprinkler systems start. Juice Box has learned his task so well, he's made it possible for her to leave home and come back again.

Actually, when she thinks about it, she says, Juice Box does more than that. His presence made things better at school too. Kristin says she's never minded getting attention — she wants to major in theater if she can get it together enough to go to college — but at school she felt exposed after she got sick. She kind of hunched over and scuttled between classes, and the longer it went on, the more she got pissed off. She added a couple of piercings to say she was okay, goddamn it, and then there was this point when she started twisting and playing with her hair and finding reasons to be alone so she could twist it until she'd pulled some out.

Then Juice Box started going to school with her. He couldn't be a bigger, hairier beacon that something's wrong with her, and at first she dreaded that, but Kristin says people at school were so interested in him that she was off the hook. They still stared, but instead of staring at her, they smiled at him, and then at her, like she was his friggin' posse — she laughs — and that became a sort of way back in to normal. Teachers lightened up too, like the dog would make everything right. How can people have so much faith? Kristin wonders. But Juice Box *does* make things better, actually. He got used to her sequence of classes. He walks a little faster to the rooms he really likes. Kristin's teachers joke they wish they had Juice Box and her in every class; students seem better behaved around him, as though subconsciously everyone wants to impress the dog. "Hey, Juice Box, hey, Kristin," people say, the way they say hey to anyone else. Kristin doesn't mind taking second billing to her dog. At the moment she prefers it.

She likes looking down to the dog that rests beside her. He's a proud dog, she says. He's my rock star. It didn't take him long to know she was his to take care of.

Kristin and her dad taught Juice Box how to find places farther from home. He was used to carrying things and picking up canes mostly, so this psych dog stuff was all new to him, but he seemed to enjoy the walks that got longer and longer and went farther and farther. There was this goal of Kristin being able to walk to school and back alone and to believe that she would get there. Kristin and her father worked to teach that to Juice Box. Somehow the twice-daily walks with the dog took some of the heat off their relationship. Kristin was less angry. Her father seemed less sad. She thinks now that Juice Box knows the words for the ten places she most commonly goes. School was the first he learned, of course, and friends' houses and her new job, where she's learning how to — *get this* — train dogs! Kristin's more independent now. She says she can think about being able to go away to college, maybe, and believe it might be possible.

Kristin describes the first time she walked out to the neighborhood park not far from her house. She and Juice Box started off to this place she knew well as a child (she has a scar from a headfirst trip down the slide when she was nine), and they walked there and back one night. Not so big a deal for most people — there were little kids there who knew their own way home in the dark — but it was the first time she'd been out alone in a long time. And at night! They got to the playground; while Juice Box watched, Kristin sat in a rubber swing and wound its chains so she could spin out of the twist a few times. She tilted back as she spun and watched the stars wheel. Then she and Juice Box walked back to their house. Not so big a deal, and at the same time, it was huge. The park is only a street away, but Kristin saw her parents standing in the doorway beneath the porch light, her mother weeping, her father beaming. They were all trying so hard to act casual. "No applause, no applause," Kristin said, laughing, as she walked up the driveway with Juice Box, but when she got to her folks, she made a big stage bow, right and left.

# 18

I DON'T OFTEN FEEL LOST, and even if lost happens, I'm not usually frightened by it. Perhaps my twenty-five years as a pilot makes some kind of difference, with its emphasis on knowing where you are or being able to figure it out if for some reason your flight goes astray. Probably the twelve years in ground search-and-rescue contribute to some confidence too. Orientation is a huge part of what we do as searchers — it's not only about finding the missing persons, but also about knowing where you are so you can get them back to safety.

The most lost I've ever felt was in a moment when I knew exactly where I was. It happened while I was piloting an airplane that had a major electrical failure as I was en route to a city several hours away. I was flying in the clouds on an instrument flight plan, no visual contact with the horizon or the ground below. The weather was overcast, heavy, sullen — solid IFR (instrument flight rules) above and below. An alternator failure had been disguised by a glitch of faulty wiring, and when the battery too began to give out, I heard the radio distort and saw the electrically driven navigation instruments begin to falter.

The controllers and I knew what was coming and had minutes enough to make a plan. These kinds of events are rare, but there are backup safety procedures. I had solid information about where I was, what terrain lay below, and the condition of the aircraft. We had physical maps. There were instruments in the airplane that were not electrical for just this reason. Nonetheless, I felt a sick little moment of thrill when radio navigation and communication failed, and I was still in an airplane ghosting forward through the clouds. Instrument flight can be very like a dream state to begin with — all that milky white above, below, and all around; the tangibles are on your instrument panel, and there's a special kind of quiet on an instrument flight when some of them go and your connection to earth falls away. It can feel a little naked, like the worst of your vulnerable dreams. A pilot can't stay naked long.

The good news: When electrical systems fail, a pilot still knows power, heading, airspeed, attitude (wings level or banked, for example), climbing or descending or straight and level — and the backup procedures for this kind of situation that have been hammered into us at every level of flight training, from beginner to advanced. Those procedures work, if the pilot adapts quickly. It takes discipline not to get spooked.

I followed those procedures, and I didn't get spooked, but I still remember that loss of radio navigation and contact, the brief surge of peril. When people who have disorientation episodes tell me what it feels like to have the known world fall away, I think back to that moment in the clouds, and I try to imagine what it would be to feel that lost not for just a moment, but for an interim that must feel like forever.

*Feels like forever* is about right, I'm told.

There are all kinds of lost, or forms of *disorientation*, to use the clinical term. A person can be disoriented in space, in time, and in person. The first two I can more immediately understand; being lost in person, much less so. Participants in mental-health forums describe these situations vividly.

There is lost in terms of location, of course, where landmark recognition and basic orientation skills are gone. What would it be like not to know your own house or arrive at your workplace and have no idea where to go next? I have searched for missing people with these kinds of disorders, people who, unable to recognize locations they should know well, simply hide wherever they find themselves or follow the path of least resistance, taking a sidewalk or street or well-trodden path wherever it leads. For those who critically wander, the inability to orient to landmarks plus the inability to measure the passage of time can be a dangerous combination.

For some, the concept of time is difficult to keep hold of. Many of us might relate to situations where time passed too quickly or crept too slowly, but we can typically recognize when something takes about an hour or that it's getting late and supper should be soon. Many of us get up at roughly the same time, feeling the hour rather than knowing it. But there are people who cannot feel the progression of a day. The nice long walk they plan to take might turn out to be a quick out-the-door-turn-around-and-back-again, while a half-hour prep for work stretches into three hours and the surprise of an angry phone call from the boss. Some can't figure out when to go to bed. Some go to bed only to get straight up again. Pots boil dry on the stove. The problem puts a strain on families and makes it difficult for a person to hold down a job. "It's not that I can't tell time," one woman tells me, "it's that sometimes I have no internal measure." She has flooded a bathroom before, starting the tub and walking into the next room for what feels like a moment and is actually more than an hour.

There's also lost in terms of identity, a dissociative state where individuals don't recognize their own faces, for example, or connect their hands to themselves, or, in some cases, have a real sense of who they are. The failure to recognize can carry over to friends and family, literally estranging the sufferer from everyone, including himself.

For some, lost can mean a terrifying mixture of two or more of these. *I don't know where I am or how long I've been here and I have no idea what to do to make this right.* That's LOST in all caps, says one commentator, a hopeless not knowing what to do in the next minute, or the next day, or for the rest of your life somewhere out there in the void.

Psych service dogs can be trained to assist those who get lost in all its variations. There are time-management dogs that can insist depressed, withdrawn partners wake up; there are dogs that can nudge slack hands until their partners recognize the hands as their own; there are dogs that, at a word, lead their panicked handlers to exits or to family they've lost in a crowd. A psych dog's location tasks not only help a handler understand where he or she is but also prevent a handler from making a dangerous move (like stepping into traffic) or help the handler get to a place that feels safe — away from crowds for example, or outside into fresh air, or back home. They are intricate tasks, skills built upon skills, tasks built on literally hundreds of training experiences that prove to a partner that his dog's assistance can be trusted.

Is Jake Piper up for these kinds of tasks? I don't know, but we'll try.

We'll start with service dog location tasks first, perhaps the easiest of the three kinds of lost. Certainly, location tasks are the most familiar to me. Such tasks represent the flip side of the job I do beside a search canine — "find our way back" rather than "take me in to find," two ends of the lost condition joined by a swath of scent. Because Jake models Puzzle, and Jake is a happy, competitive learner, we bring Puzzle into the training process — a demo dog for the student dog — and I am behind them and beside them, taking notes and learning fast.

When Puzzle works the Home or Take Me Back command, often Jake Piper goes along. His devotion to the golden seems to make him learn faster. I'm eager to see how much of what she's doing he picks up just by proximity. On his own, since Jake Piper has no orientation or nose-work background at all, I decide to start him with a simpler orientation task: Door. For the human partner needing to find his way out of a building, *Door* means "Help me leave this place," "Take me to the fresh-air source," or, as one partner puts it, "Get me the hell out of Dodge."

Training one dog in a houseful of others is always a challenge. Training one dog in a houseful of others during a rainy late autumn even more so. Jake needs to learn the Door command. Puzzle can certainly learn it too (why not?), but I didn't really have it in mind to train the Pomeranians also. Little center-stage creatures that they are, they grumble every time they are sidelined. Maybe they are up for the challenge and maybe they strongly sense that I have underestimated the depth of their capacities, or maybe it's because they know there are treats involved with every success, and they are all food hounds, and this exclusion is seriously inappropriate. I am used to their protests. Many a Pomeranian glared at me through the window when Puzzle was first doing search exercises in our backyard, six years ago. Now the Poms have the reverse condition. On sunny days, they are led outside when the good stuff is going to happen in the house! As we set to work, I can hear their fussy voices at the back door.

Jake, Puzzle, and I start by defining the word *door*. "Door," I say, and I stand right by it, and when they come to the door and sit at my feet, they each get a puppy biscuit. This takes a couple of rounds, and then I stand farther down the hallway and say, "Door," and first they come to me for a treat, but in fairly quick order the dogs recognize that's not what's wanted. They need to go to the door and sit, and they do. They get treats for that too. This takes about two fifteen-minute training sessions to become reliable.

It's important that the dogs don't block the door on this command, so now that they recognize that for the moment, the word *door* means our front door, I reward them only after they sit to the side of the door, so that it can be opened, rather than when they block it. This was already in Puzzle's skill set — she's had a lot of experience with public spaces and swinging doors. Jake seems proud of himself when he gets it. He also seems to have a sense of the spatial and the practical, recognizing that if I'm going to open the door, it's easier if he's not in the way. After another day of short training sessions, he dashes down the hall to sit beside the door, one paw slightly lifted and posed just so, beaming at me like a Chinese foo dog.

Now it's time to broaden their understanding of the word. I'd like the dogs to lead me to the *closest* exit to the outside, so I need to teach them that the *door* can mean the back door too. We try this a few times from positions in the house that are closer to the back door than the front. For both Jake Piper and Puzzle, learning that *door* means any exit and, ideally, the closest one to where we are, takes some doing. At first, they rush happily for the front door no matter where we are. So I start the process all over again at the back of the house. I say, "Door," and move to the back door instead of the front. On the first attempt, I have forgotten that all the Pomeranians are outside on the deck, and when Jake and Puzzle head for the glass back door successfully and get treats for it, the Poms see the endgame from outside. They stretch up on their back feet and press their angry faces to the glass. We ignore them and train on.

It takes a weekend of short training sessions to get Puzzle and Jake to hear the command and make a choice of door based on location.

"Remember that the closest exit may be behind you!" I call to them, imitating a flight attendant's briefing.

When I give the Door command just inside the doorway to my bedroom, the dogs are truly torn. I watch them with interest. They dash into the corridor. Jake, who can scramble in reverse somehow, moves backward with his eyes on me. They seem stymied for the moment. The back door and the front door are about equidistant. But the front door, visible from the corridor, wins. Puzzle makes the choice, Jake just behind her, both of them glancing over their shoulders to make sure I am following.

The next step for them in the task is, of course, partner loyalty. It's not enough for the dogs to find the door. They need to make sure I find it too. There are all kinds of similarities here to a search scenario, where the dog must not only find the victim but make sure the handler knows he's found him, and in some cases lead the handler back to the victim. I expect Puzzle to catch on to the partner-loyalty concept pretty quickly. She does, but her loyalty is of the search-dog kind. She steps toward the middle of the corridor, *wroo-wroo*s at me, then turns back for the door. This is the "follow" signal we've had in the search field since she was a puppy. It's clear between us there, and she expects it to be clear between us here too.

Jake Piper is more unpredictable. Food-driven Jake might be so focused on the treat that he fails to recognize he receives it only if I actually get to the door. We try several training sessions where I move more and more slowly, with greater difficulty, and Jake, with his generous eye gaze and his urgency, seems to get it. He knits great stitches between me and the door — trotting back and forth and back and forth — until I get there.

The Door command is going pretty well. Puzzle and Jake make the training fun. *Wroo-wroo-woo*, Puzzle says when she knows she's got it right, her golden brag language, while Jake wiggles beside the door, waiting for the praise he thinks he's earned.

Then the rainy season truly hits, and separating the training dogs and the Poms gets a little more difficult. I try putting the little dogs in a bedroom so that Jake and Puzzle can concentrate (and maybe even hear me) in the rest of the house. All that distraction is a good test of the working dogs' focus. On the other side of any door, the Poms are completely aware of what's going on in the living room, the study, the kitchen, the hallway. I hear them put their noses to the crack beneath the door and huff. Occasionally one of them throws his entire body weight against the door, like cops in those crime dramas who shoulder into an apartment full of bad guys. Every once in a while, Mr. Sprits'l yaps an *augh!* I can hear the tap of his feet on the wood floor, and I know he's spinning, one *augh!* per 360-degree rotation.

After a week or so, Puzzle and Jake have clearly got it. *Door* now means "the-closest-exit-outside-no-matter-where-we-are." *Door* also means "and-make-sure-the-human-gets-there-too." I feel a little bad about the Poms, the tragic little overlooked, underestimated Poms, and now that Jake and Puz seem assured about the command, I decide to invite any Pomeranian that wants in on the action to have a go.

We'll have a little fun. "Door," I say in my bedroom, armed with a pocketful of treats. Jake and Puzzle race to the back door and sit, and I follow them readily, but the Poms at first follow me, because I have the treats. I start with them the way I started with Jake and Puzzle. *Door* means a treat when you get there, not before. A couple of them (Jack and Smokey) figure it out quickly and are happy to run to the door and sit for the treat. One of them (Mr. Sprits'l) would rather scold me from ankle level all the way there. One of them (Mizzen) is a natural. She races to the door and back to me again, there and back to me again, there and back. *Hoor!* she says, tap-dancing across the wood. She can get to the door and seems to know what the word means, but it's all so exciting she can hardly contain herself. *Hoor! Here's the door! Aren't you here yet? Hoor! Let me come back to you! Hey! Look! Over here! Hoor! Here's the door!* She is thrilled with Door. She is thrilled with the knowing. She is thrilled with the treats. Mizzen-monkey makes me a little dizzy.

We shape the command. Ultimately, the door dogs, the ones who get there, make sure I get there and then sit out of the way, and they get the treats. The recalcitrant dogs eventually get treats too, but it takes a little longer with them tripping over their personalities along the way.

The quicker dogs get impatient with the slower ones. At one point after the Door command, I notice Fo'c'sle Jack quivering at the door for his treat, but I'm having a tough time making my way there. Mr. Sprits'l is slowing me down, making figure eights around my feet as I walk. About three yards from the door, I catch Fo'c'sle Jack's eye. He glances at me, then looks at Mr. Sprits'l and sighs.

There comes a day, not quickly, when all the dogs can find the closest door when I give the command. They rush to the front or back door as a herd of ears and tails. A few of them come back for me. Some of them don't. Then they quibble with one another for sitting position. Knowing that it's easy for dogs to learn our habits like dance moves, I try to vary when and where I give the command. Sometimes they're awake, sometimes some of them are sleeping, but "Door!" I'll call, and I'll hear the mutter and scramble. It is the doggy version of my search pager going off, that sudden call to action, and I have to laugh at their fumbling *Huh? What? Oh!* as they stampede through the house to find me and then lead me to the closest door.

But they seem to enjoy it. All of them — young and old, fast and slow, strong or a little bit frail. "Door," I say, and they find me to guide me. Mr. Sprits'l seems particularly glad to tell me where to go.

# 19

"ROSCOE PLAYED BALL the way Cal Ripken played ball," says Alex. "He took the game very seriously." Alex remembers how his off-duty service dog became famous at the dog park, an unexpected runner among plenty of other running breeds. Roscoe's athleticism attracted attention.

Alex speaks of his partner of four years with a kind of wonder. Like so many rescuers, he brought one dog home, you see, and he ended up with another. Alex never regretted it. There is still not a day he doesn't think of that boy, not a day he doesn't feel the dog's influence in his own life and recognize how Roscoe changed it. Homeless twice over, Roscoe became a rescue success story. The same could be said of Alex.

Alex was a shelter volunteer when he met the dog that would become his partner. Roscoe was a longtime resident there, the kind of dog in a shelter that attracts both attention and apprehension. A light brown pit bull–whippet mix with a generous smile and a white blaze down his chest, he was a favorite of the shelter staff — friendly, handsome, and personable. He had plenty of time to become a staff favorite too. He was difficult to find a home for. Roscoe had both size and the smoke of breed bias that surrounds any dog that looks like a pit bull.

He was lucky in his placement; the Linwood, Washington, Progressive Animal Welfare Society (PAWS) that housed him euthanized dogs only in cases of extreme poor health or behavioral issues. Roscoe had neither of those. At PAWS, Roscoe was given time to find a home, and Alex says the dog thrived among friends at the shelter, a place where he had known only kindness. Volunteers, Alex among them, paid him a great deal of attention.

Alex saw something special in Roscoe. There were so many good things about him, but he remembers that Roscoe was a dog that would look you in the eye and concentrate, as if, if you gave him a minute, he could understand. This dog paid attention. He wanted to connect.

Alex needed an assistance dog. He had an eye on this dog as a service partner as the documentation necessary to support his need finalized, but before Alex's paperwork came through, Roscoe found another home. Alex couldn't help but be disappointed at the same time he was glad. A young couple had taken him, and it appeared that Roscoe's story would have a happy ending, never a certainty for any dog, and especially not for homeless dogs like Roscoe, too often bypassed.

But a few months later, Roscoe was back at the shelter. The loving couple that had adopted him had also abandoned him, moving away and leaving him tied up in their former backyard — a hint at the life they had failed to give him. When neighbors realized the couple wasn't coming back, they fed Roscoe until animal control stepped in. AC found Roscoe's microchip, the microchip that still contained PAWS information, and what could have been a terrible last chapter turned out well: suddenly, the big dog was back at the shelter among friends. He seemed relieved, Alex says. This was where Roscoe had known safety and happiness, and though there were no signs of abuse by his former owners, Roscoe was clearly the victim of neglect. He came back anxious and undernourished. He would not be neglected at PAWS.

He would also not be at PAWS long. The mental health documentation necessary for Alex to have a dog in no-pet housing had come through in the interim, and Alex, who needed a service partner and was determined to owner-train *and* to rescue, knew the moment Roscoe P. Whippet came back to the shelter that this was the dog that could make a difference. Alex knew also he was a man who would return the favor.

His condition might surprise most people, Alex says, because on paper it looks like everything's great. Long involved with Toastmasters International, a nonprofit public speaking and leadership group, Alex is an extrovert. He is comfortable and confident in front of crowds. Alex has done theater and even some standup comedy.

But there are issues. One-to-one interaction is different than public speaking. Alex wrangles attention deficit disorder, focus problems, and a social-anxiety disorder that makes even some of the basics of daily living difficult. In the days before Roscoe, at its worst, the condition was agonizing. Job interviews, first dates, meeting strangers in any context could cause intense anxiety attacks. Alex says there were times when even small, simple interactions like asking for directions or for help in a store were almost impossible. Self-consciousness overcame him — Alex couldn't meet people's eyes; he sometimes couldn't speak at all.

His job could well have been part of the problem. Alex worked on the distribution end of his own book business and spent the majority of his days alone in a warehouse, sorting books and working online, rarely seeing or speaking to anyone. It was a job he loved in a business he was passionate about, but the less he interacted with others, the more difficult it became to do so. Attention and focus issues gradually became problems at the warehouse, issues that had also kept him from driving. His separate problems added up to a whole, and there he was, a withdrawn, center-stage extrovert tipping the Myers-Briggs scale in one direction while living an anxious, largely interior life that pointed to another.

Alex is smart and self-aware. He could feel his encroaching anxieties, the sense of his own withdrawal. The more he was alone, the more he felt himself avoiding strangers and unfamiliar situations he would have once faced; the more he was alone, the more he was aware of the division in his public and private selves.

Alex was as startled as his audience the day he broke down during an early-morning Toastmasters address. Toastmasters is famous for its supportive community, and Alex knew this group well. He was comfortable there. He had been involved long enough that he'd become the chapter's president. The day's exercise was not meant to hold surprises; like most other Toastmasters events, there was a speaking topic for the meeting, and each member would get up to address the group.

Pets and Animals was the subject of the day. There were certainly more difficult topics, and Alex remembers that the subject didn't seem particularly volatile when they'd come up with it. He enjoyed animals, had had a few dogs in his childhood, but he had not had a pet as an adult. No problem, he thought. He liked animals well enough that he knew he'd find something to say. Other members got up to speak of their own pets past and present, and when it was Alex's turn to speak, to close the meeting, he began to speak but faltered, unable to go on, and then wept hard enough that the vice president of the group had to take over. Twenty or so Toastmasters members on folding chairs in a church's Sunday-school classroom: it was a routine he enjoyed; he was familiar with this space; he was confident with these people — and yet somehow he got blindsided by a topic that might have been given to first-graders.

What had happened? Alex isn't sure. Perhaps it was the collective power of the other speakers' stories, some joyful, some meditative, some sad. Perhaps he suddenly realized the depth of his own disconnect. Whatever it was, he felt a dark space widen, and there it was, this grief he wasn't prepared for.

The situation was relieved only when another member of the group came up to talk to him about PAWS, where she had once volunteered. Walking dogs, socializing dogs, comforting and caring for them — she suggested that if Alex was not able to have a dog in the home he was renting, this might be enough in the interim, this ability to be with some dogs and do some good.

That public-speaking breakdown was a ferocious disclosure — it was the stuff of nightmares for many people — but Alex has no regrets. Without it, he might never have heard of PAWS and almost certainly would never have volunteered there, a good move in so many directions. His interaction was hesitant initially, but at the shelter he became more engaged, communicative — first with dogs and then increasingly with the other volunteers — and his quality of life improved. It was progress by inches. While he was waiting to receive professional documentation of his need for a service partner, Alex was searching among the rescues for a dog that might fill that need. Moved by their homeless numbers, Alex knew he wanted a pit bull or pit bull mix. He got that in skinny, abandoned Roscoe.

He also got something else.

*I adopted a pit bull and brought home a whippet* — that's Alex's take on Roscoe now. He was a good dog, a great dog, even, but he was a dog that needed space, and he was a dog built for speed. His energy had been obvious in the shelter environment and Alex thought he "got" the dog that Roscoe was, but he was in no way prepared for the dog that Roscoe would become. Now safe and well cared for in a loving environment, with an owner that paid attention, the scrawny pit bull mix quickly matured into a powerhouse athlete that lived to run.

The first service the dog gave Alex was indirect, less by task than by change of habit. Roscoe needed a large life. When the dog lived too contained, he grew anxious, restless, chewed at a lick granuloma on his foot. There was no way to restrict a creature like Roscoe to a low-key life as a house dog. Alex knew it would be an impossible cruelty. He was intrigued by his changing sense of the dog he'd brought home. Had Roscoe's energy been the deal breaker for the couple that first adopted him? The dog needed dedicated outings and room to run, and he needed wilderness. Responsive and responsible to his dog, Alex was shoved from seclusion by sheer force of nature. Roscoe could not be denied.

In the wilderness, he became a different dog. Sure, he blew off plenty of energy playing ball the way any dog might, but then he wanted to explore — anything, everything, winding through gates and along trails, choosing rough terrain over easy. He was extremely visual. Where possible, Roscoe would scale the face of cliffs to reach the higher view, Alex scaling and viewing behind him. For many service dogs, this amount of drive would have been a problem, but Alex says Roscoe, who got him out of the house and thrust him into situations where he'd have to interact, was exactly the dog he needed. Alex could not remain a recluse. Roscoe's service came wrapped in a certain tough love.

Roscoe brought other issues to surface. When bus schedules didn't match up well to their joint needs, Alex overcame his reluctance to drive. He's up front about this: "I learned to drive for my dog." His concentration and attention had improved enough for him to feel safe in a car, and with Roscoe in it, his sense of responsibility for Roscoe's safety helped him focus. So he and the dog loaded up daily in a two-door white Mazda. Sitting behind the driver's seat, blissed out by speed and weather, the dog loved to lift his face out the window to the rain or snow; he loved to be on the highway going fast.

Roscoe's idiosyncrasies were part of his charm. Photographs of him seem to show an all-terrain, all-weather kind of dog devoted mostly to his ball, but Alex describes a conflicted, almost dainty way about him at home. On the move, Roscoe never worried about mud, water, or sludge of any kind — he braved it all and once ran up to Alex so muddy that Alex didn't recognize his own dog. But at rest, Roscoe would sit with his feet well up under him, never fully down; he disliked having his bottom touch the ground. How could a dog be so fastidious at home — the word *prissy* comes to mind — and so gregarious in the woods?

Roscoe could also obsess over things, wanting *this* ball, *only this* ball. He would ignore all others. Sometimes on a given day he would return a fetch, no matter which direction the ball was thrown, by taking only one track back through the woods. Alex liked to test this. He sometimes threw the ball in directions that would make it so much easier to come back another way, only to watch Roscoe take the harder route over rocks and through brush to return along the favored path of the moment.

He didn't mind taking center stage. Where some dogs might have fought it, and Alex had expected Roscoe to resist, Roscoe graciously acquiesced to a Flash Gordon costume for a Halloween dog-park event. In fact, Roscoe not only acquiesced but seemed to enjoy the attention he got blazing back and forth through the crowds in red-and-yellow polyester with lightning bolts bobbing from the back of his head.

Alex laughs and says that to some degree, he and his dog were maybe too much alike. They had the same kinds of issues, which shouldn't have worked for a dog in assistance to Alex, but somehow it did. Perhaps it was because Roscoe was more attuned to humans than his own kind. Rescued Roscoe had rarely played with groups of other dogs, and now the loose social order of the dog park confused him. He was excited by the noise, the energetic rabble, but didn't seem to understand how to engage with it. Roscoe ran with a pack in periphery, circling wide, rarely in the midst. Alex could be describing his own nature in a group of strangers, he says. Not unfriendly, just self-conscious and extremely uncertain. Dog-to-dog social skills came to Roscoe by way of Lulubelle, owned by Alex's housemate, Dan. Lulubelle, herself a brindle pit bull mix, was aloof with most dogs but immediately adored Roscoe when he came into the house. She was an affectionate, timely influence. Her presence seemed to lessen any stress Roscoe had. Alex jokes that while Roscoe was his service dog, Lulubelle was Roscoe's.

Smarts to burn. Roscoe had plenty of smarts, but he was an odd dog at first. He wasn't sure what to make of fetch. Roscoe was simply too infatuated with the complete perfection of ball. Initially the dog would simply go after the ball and then run with it in his mouth, joyful, heedless of everything and everyone else: *Here is this thing I've caught and now I have it I have it I have it.* Fetch took a little patience on Alex's part. Plenty of games began and ended with just the one throw. But Alex could see the dog was eager and conflicted. He wanted to possess, but he also wanted to run, and he wanted Alex's attention too. How to do all at once? Alex taught him to fetch by alternately throwing two balls. Once he learned the routine, Roscoe seemed overjoyed by the concept of a game involving all of his favorite things: the ball, the run, and his person, all of which could be had over and over again.

Roscoe too was attentive, so ardent a communicator it seemed like he was almost straining to be human. Where he was remote with other dogs, he was direct with humans — a lot of interaction, a lot of eye gaze, a consistent use of the common nonverbal language he and Alex both understood. He had some smart-dog tricks. Alex could put photographs of Roscoe's toys in a low place on the refrigerator, and Roscoe would point to the toy he wanted. The dog listened, too, and remembered what he heard. Over time he understood far more words than Alex was prepared for, and when planning a day's activity with his housemate, Alex found himself creating synonyms for *park, ball, play,* and even for Roscoe's name — the dog grasped the essence of even long sentences too well, and he would not be patient about a dog-park outing that might be five hours away.

How well did Roscoe understand Alex's condition? Alex can't be sure, but Roscoe was "one hundred percent my dog," Alex says. "He was *with me.*" The dog had a strong sense of when to lead Alex out of places and when to intercede where he was. Serving somewhere in the netherworld between emotional support animal and assistance dog, Roscoe helped Alex build the skills that had previously eluded him. Think what a dog needs from a thoughtful owner: attention, interaction, considered care, and socialization. These were the things Alex also needed — and needed to be able to give.

Alex says at first a few friends were skeptical about his documented need for Roscoe, but later they admitted they could see the difference. Alex's improvement beside the dog was tangible. The dog's solid presence translated to a sense of strength for his partner and made it possible for Alex to hold conversations with strangers. It held the panic attacks at bay. Talking about dogs at the park came fairly easily to Alex, and from those small interactions came familiarity, confidence, and control, sensations that gradually stretched beyond the boundaries of the dog park and became more reliable in daily activities. Alex grew even more outgoing. His volunteerism expanded from PAWS to a wildlife rehabilitation group. In time, Alex didn't need Roscoe beside him everywhere he went, but Roscoe's influence traveled.

Alex credits part of his own growth to his love for Roscoe and his willingness to give back to the dog. Roscoe's commitment was a gift that had changed him. Focusing on Roscoe's needs had also changed him — preventing Alex from thinking too much about himself.

When he got sick, Roscoe was five years old, young by most dog standards. The symptoms were vague, if they were really symptoms at all, and in a large-living dog like Roscoe, they were difficult to see. He began to slow down a little in 2010, Alex says, taking shorter runs on some days and resting more often, sometimes in the middle of trails he would once have taken at speed, but the change was so gradual that at first Alex attributed it to approaching middle age. Dogs mature. They slow down.

But when a lump on Roscoe's jaw that appeared to be an abscess was diagnosed as osteosarcoma, everything, for a moment, stopped. Osteosarcoma, or bone cancer, is an aggressive and painful disease that in dogs often shows up in the limbs — near the shoulders or knees. By the time of diagnosis, an estimated 90 percent of cases have already metastasized. Timely amputation of the affected limb can save the dog's life, but amputation was not an option for Roscoe. The disease moves quickly. He was given three months.

In the wake of that prognosis, Alex and his housemate investigated every available treatment option for Roscoe, including chemotherapy at special canine cancer centers, dietary supplements, naturopathic injections. There were plenty of options, many of them expensive, some of them far-fetched, but few with real hope. The consensus suggested that this disease was a bastard and that chemotherapy might buy Roscoe a little more time but at the cost of either medicated stupor or extreme discomfort. Alex wanted neither for his dog. Determined to give Roscoe quality of life for as long as possible, Alex and Dan decided to take on whatever debts might come. Ultimately, Alex opted for naturopathic treatments offered by Roscoe's own vet.

Alex's mind was on his dog's happiness. Roscoe deserved diligent care in that sense too. Despite the time now given over to medical treatments, all the dog's favorite rituals remained, were even extended, when possible — the daily dog-park trips, the games of ball, the tough trail walks that Roscoe continued to choose even as his condition worsened. Roscoe loved the wilderness and he loved his rituals, Alex says; he had deep attachments to routes and places and situations. He was a dog that expanded outdoors, and sometimes, in the early stages of osteosarcoma, it almost seemed he could outrun this; his heart was wild. When Roscoe could no longer run, he loped. When he could no longer lope, he walked. He rested in the shade more and more often.

The cancer spread through Roscoe's jaw, and games of ball became impossible. Alex marks that transition at the day when Roscoe could no longer hold the ball in his mouth. Adaptable as he was driven, Roscoe wanted his game. He wanted something thrown. He chose a stick; Alex threw it, and instead of picking it up to return it, Roscoe ran to the stick and placed a paw on it, waiting for Alex to catch up and throw it again. Fetch transitioned to a sort of relay race. Almost as beloved, Stick would be the new game until both the game and the trail were too hard on the weakening dog.

In time, Roscoe surrendered play in favor of simply going to the park to watch the other dogs, a choice that surprised Alex. Roscoe had never really joined the thick of pack, but now he enjoyed watching the distant noise and tangle. That change resonated for those who had known Roscoe at full speed. Dog-park regulars — human and canine — quietly visited them. For Alex, the former recluse, the company was welcome. Camaraderie had now become second nature. The other dogs seemed to be aware of Roscoe's illness. They sniffed, circled, wagged small, acknowledging him gently.

Alex remembers Roscoe in transition, the dog's considered movements, the way their relationship slowed. They had been fast once. They had blazed trails. Quiet companionship now replaced that exuberant dash. In his final weeks, Roscoe needed Alex to be near and to be still, and Alex, for whom stillness had never been easy, found it in himself to give. Alex remembers long hours at the edge of woods beside his partner, the dog with his wounded head lifted, taking in the wind.

# 20

JAKE PIPER'S MAKING SLOW sense of the Home command. I'm
not sure how much he's really got it. He follows Puzzle readily when
she leads us home, but I notice that he is clearly *following* her. He is
a moment behind her dance steps, always, and doesn't appear to be
making any choice of direction. When I give him a handful of trial runs
beside her, I realize that as long as she is with us, he will be content to
let her lead. I don't think he's got it. It's time to work him on the Home
command alone.

I've watched him work his nose since the very first day he came to
us, and what I know about Jake Piper is that he's got a keen sense of
scent and a high drive impulse. But, unlike Puzzle, he doesn't have
six years of finding a specific something and leading me to it with his
nose. Jake also naturally works more head down across the turf than
head up across the wind. He can trail a rabbit's recent path through
our backyard quite easily, but it's Puzzle, head up, who seems to snag
the airborne scent of passing humans long before they reach the house.
It's Puzzle that picks up on the roof-hugging squirrel pressed flat to the
tiles above us. Of course, she is a field dog by birth and long trained to
work air scent, and he is all raw nose talent and completely unversed.
I'm interested to see just how quickly he picks up the Home command
and recognizes what I need him to do. I'm curious if he'll consistently
backtrack our trail or if he, too, in time will simply take us home by
moving from scent zone to scent zone regardless of the path we took
outbound. It's possible he won't pick up this command at all.

Jake learns words quickly, so I start with teaching him what I mean
by *home*. Walking along the boundary of the property, I'll suggest we

go home, and then, as we approach the front of the house, tell him to find the door. Jake has successfully learned that the Door command can mean door *in* as well as door out, so I hope to build on that understanding. Where once it was just about finding the door, Jake's task now demands he find home and the door. For a week's worth of sessions, I simply say, "Let's go home," as we approach the house, adding the Door command as we step onto the property. Jake learns commands well. The first time I say, "Let's go home," along the back fence of the property, and he chooses to run the length of the fence and then turn right to get to the front door, I mark it as a success. Yes, he did pee over other dog marks on the way — a quick hike of leg out of form rather than function — but he got us there. He enjoys the command, the job, and the big, big praise for a good dog doing well. In this he is much like Puzzle. Home is a happy command to give a once abandoned dog like Jake. Every time we work it, I'm reminded that in a way Home celebrates what he nearly never had.

We begin to train farther away. This is tougher for Jake, working across a universe of distractions. Even one house down from ours there are enticements: a cat arching in a window, any number of piss marks on trees, a child's sock, a dead pigeon. Before we can nail the Home command, Jake has to reliably Leave It, a term he now understands. Usually good about it, he occasionally plays dumb and lunges for whatever (*Never heard that command in my life*). He also plays deaf (*Even if I do know the command, I didn't hear it*).

Jake's a curious beast, leaving the taunting cat and the rotting pigeon much more readily than the piss marks, leaving the child's sock most reluctantly of all. He doesn't try to snatch the sock as toy, but he's curious about it. When Jake does Leave It on the second command, he looks up at me in puzzled innocence, as though he's wounded by my tone of voice.

In a few days, from one house away, he leads me home. In a week, from two houses away, he leads me home. When the month is out, I can give him the Home command as we round any block leading to the house, and he'll take me there, long lead drooping and scraping across the sidewalk, a loose connection between us so that I can be certain I'm not cuing him with tugs even I don't recognize. For a time we work into the sun so that my shadow is thrown behind me — I want to make sure I'm not even cuing him by some lean of body he can see, though I'm not sure he recognizes what a shadow is.

We have several good Home finds from a block away, once even approaching from a side of the block we had not taken outbound, and I think it's time to let Jake Piper advance a little more. We take a long, free-to-be-dog walk into town, and halfway back I put on his service vest. He stands still for the putting-on-of-uniform, slides into it easily, and I see the change of demeanor I see in Puzzle when the vest goes on, as if he understands which rules apply.

"Take me home, Jake," I say. We are about three blocks away. It's a big step up from the block he had been doing, but we're on the very road we took outbound, and we are walking into the wind. With any luck, it's blowing straight over the house and into our faces. With any luck, Jake has so much of our outbound scent and home's scent that the path back glows.

Jake perks at the command and starts off with great energy. Too much energy. For a moment I have to rush to keep the lead slack between us. If we weren't near traffic, I'd drop the lead entirely to see which route he'd take. Jake's head is lifted. The spotted left ear is standing almost straight up. The right ear twists like a corn chip. Everything about him looks happy, and with a terrier's easy, distinctive trot, he moves confidently in the right direction. His pace is steady: Jake-Home-Jake-Home-Jake-Home-Jake-Home. A couple of times he turns around to shoot me a glance. It's a check-in but so confident and prideful that it reads less like *How am I doing?* and more like *I am so on this. Who's the good Jakey?*

*He's the good Jakey,* I think, and I am just about to share his overconfidence when suddenly he shivers all over, the nose drops, and the tail goes from a sway to a wag. This is the very kind of animation we may see in search dogs the moment they catch human scent. There is nothing about the Home command that should torque Jake up in this way; I'm thinking that even as he moves from the trot to a scramble and, nose down, begins to pull me along the sidewalk — *right-direction-right-direction-yes-it's-the-way-we-came* — and then suddenly goes across the street on a diagonal, onto the opposite sidewalk, and into a row of bushes. Jake is crittering. He has chosen a place where possums like to sleep off the day, and I have no doubt he's trailing one now. He is on it. He thrusts into the green so deeply that all I can see is his madly waving tail. By this time, I've abandoned my role as observer of the process and am pulling him into me, heaving hand over hand down the long lead. We meet somewhere in the middle of the thicket, and as a huddle of little possums scatter in the underbrush, Jake sits and turns to look at me. His expression isn't guilty. He beams as though he thinks he's done the job.

"Jake, come out of here," I say, leading him out, giving an embarrassed little wave to an elderly man grinning from a neighboring porch.

No treat. Sit for Jake. Deep breath. Let's try this again.

"Jake, take me home."

Jake stands and looks pointedly at the hedge, then back to me.

"No, Jake. Take me home."

His expression mystified, as though he can't imagine why a hedge full of baby possums is not the thing I want, Jake begins again. We move away from the hedge in a cloud of Leave Its; Jake crabs sideways, looking back toward possum land as long as he possibly can. Somehow, though, when we hit the sidewalk, he seems to shake off the hedge's allure. He is service dog in training again, and I am hopeful that this time he's got it. He's slowed from a trot to a walk, a better pace for a human handler behind him, and I'm glad to see he made that choice. (I hope he made it out of consideration for his partner, but he may have made it because we're now upwind of possum.) The lead between us is slack.

Two blocks from home. A block and a half from home, and we're still moving in the right direction, at a partner-sensible pace. A car passes, honks lightly, its driver smiling at the two of us. I smile back, because really, we are on it this time, and here we are, woman and dog, the picture of obedience and collaboration. A nice walk now; *Jake-goes-home-now-Jake-goes-home-now*. No pull, no tension on the lead, still moving in the right direction.

A block from home, Jake's head pops up with interest, and for a moment I fear he's on another possum. But no — though Jake has snagged some happy scent on the wind, he doesn't break stride for it. Instead, he moves forward with great confidence, leading us down the last block, up a low set of steps, and to the front gate. He sits and grins at me, eyeing the treat bag.

This was right on so many levels. Right direction, right pace, no startle at the car honk, a lead up the steps, to the gate, and the happy *Here we are!* Sit. The only problem: We're at the wrong house. Not only the wrong house, but one of the most splendid historic properties in our little town. A beautiful full-blown Victorian mansion with turrets and wraparound porches, its lovely landscaping bounded by a wrought-iron fence.

To Jake's credit, we passed this house outbound, and today it's immediately downwind of our own. I can see home from where we stand. So could Jake if he were looking for it, but at the moment he's happy with the house he's found for us. Holding his Sit, he perks every time I look at him. *This is home, isn't it?* I give him credit for having good taste, if not accuracy. He seems pretty sure we should just head on through the gate.

# 21

SHE HAS DREAMS SHE CAN'T wake from and rooms she's unable to leave, and she recognizes that these events, a part of her existence for two decades, are congruent to the life she lived as a child. That she still pays a price for childhood trauma doesn't surprise her. That it took thirty years for it to surface in the first place does. Nancy is a fifty-five-year-old wife and mother of a daughter who is grown and gone. She's a petite, pretty woman with a rich voice reminiscent of Big Band singers from the time she was born, a voice that tightens to a whisper every four minutes, literally like clockwork, caused by a vagus nerve stimulator implanted five years ago. The VNS is another in a long list of experimental protocols targeting Nancy's diagnosed treatment-resistant depression. It is a condition that has otherwise resisted all medications and shock treatments. The vagus nerve stimulator doesn't hurt, but when it pulses, her voice binds as though she's on the verge of tears.

She is not on the verge of tears. Nancy speaks of her psychological state frankly, the way one would speak of a house inhabited for a long time. These are the rooms you'd expect and the strange spaces you wouldn't. This is where the foundation's wonky. And God knows what exactly is going on here.

I've known Nancy for more than a decade. As happened with Paula, our slim but affectionate connection was forged online, this time through a Pomeranian message board on AOL. When we both found ourselves on Facebook years after our days on AOL, Nancy and I reconnected, finding — or looking for — other old friends too. Some were missing, never to return. Nancy remembered our friend Erin and had followed her journey beside Smokey and Misty with compassion. Much has been written about the peril and falsity of online friendships, but many of us can vouch for the support system we find there at three in the morning when a child, a dog, or a jobless spouse takes a turn for the worse. We keep up with each other, and it's not too much to say that when one of us has a serious loss, the other genuinely grieves. And now Nancy comes to me as her own path is turning.

She sends a private note online. She is straightforward. She's struggled with mental illness awhile, has run the gauntlet of meds and therapies and alternative therapies, and at the point when she felt like giving up was the only option, her therapist suggested a psych service dog.

A psych service dog? Nancy loves dogs, of course, but this is new ground. This would be a dog like no other. A dog to be *present*, a dog to intervene. She has no idea where one might be found. Nancy asks if I can help. She's excited, uncertain, but she's intrigued by the possibilities, and there's a lot of good energy in that.

She is a good wife, mostly, she says, and a better mother than she was for the longest time, when she almost derailed her own daughter in the way she too was derailed back in the sixties, before anyone really knew how to help a family like hers — five children in a household reared by a mother who was manic-depressive before the term was widely known.

Her father couldn't take it. He fled her mother's "spells" the year Nancy's youngest brother was born, and in a Catholic family where divorce was a disgrace and appearances were sacred, Mary carried on, working to support her children, drinking as an answer to her undiagnosed condition, and carefully separating the children for the sake of decency: two of the girls in one bedroom, the two boys in another, and first-grade Nancy in the same room as her mother.

Seven-year-old Nancy accidentally discovered her mother after two separate suicide attempts. Mary's recovery was a patch-and-go affair. Nancy became the gatekeeper who protected her mother from the other children's squabbles; the caregiver who got her through weeping spells; the sprinter for the phone or the long-distance runner when help was needed. There was nothing right or fair about any of it. The family was so good at carrying on that the neighbors never knew. Nancy rethinks this. How could they not have known? She laughs a little, short and sharp, in the telling. Her story belongs on a therapist's couch she says — another bad mother story, get in line. But for Nancy, the problem with talking about her mother is that somewhere in all her meticulous, elementary-school-age caregiving history, something so bad happened to her that she can't remember it, and her brothers and sisters, shuffled off to relatives in secrecy at the time, don't know enough about it to tell her and set the matter to rest.

What Nancy does know is that she was abducted during a walk home from school at age ten — a walk already made frightening by a neighbor's two large white dogs that rushed toward her and savaged every day when she passed, slamming the fence with their forepaws and hating her with hard eyes. Nancy had been told how she was to walk home from school. She was a rule follower and could not cross the street to avoid them. She was a regular, predictable figure at the same time every school day. Nancy remembers a little of that particular walk home the day of her abduction, remembers her arm being jerked and seeing a man's face before a white bag, a pillowcase maybe, was thrown over her head. She remembers also a flash of white truck. And then nothing, nothing at all, until she woke in a hospital, her mother at her bedside and her father returned to the family. She was the center of attention in that hospital ward, the queen bee, the recipient of many presents — even a training bra, a status symbol among her friends and something she absolutely didn't need — and for reasons she couldn't remember and didn't understand, she got everything she wanted and then some.

Funny, she says, that the color white figures so frequently in these memories: white bag, white truck, white bra, white dogs rushing to a fence.

Whatever had happened to her made things better at home for a time, until the police presence faded, her siblings returned, and her father left again. No one mentioned Nancy's abduction, even after their mother died unexpectedly, ten years later. As an adult working through mental illness now, she has plenty of questions and very few answers. Those who knew everything about what happened are dead, and the siblings sent off to relatives hardly remember it at all.

In the spirit of maintaining appearances, young-adult Nancy carried on, trailing the weight of her grownup childhood. Pregnant, married, divorced, scattered, she became a liberated variant of the mother she had cared for. She was in her thirties and thriving in a high-stress career with a worldwide shipping company when she had a catastrophic nervous breakdown and depression so severe she was hospitalized. Variously diagnosed — borderline personality disorder, bipolar, PTSD, manic disorder, agoraphobia — and even more variously treated with medications and electroconvulsive therapy, Nancy was ultimately unable to return to work and unable to care for her family. Depressed, agoraphobic periods were followed by manic episodes that led her to spend money wildly and disappear, sometimes for days.

Nancy has "a husband in a thousand, who stepped up," she says. Twenty years her senior, second husband, Harv, raised Nancy's daughter, maintained the household, loved and cared for Nancy with the numerous small adaptations that a marriage built on a fault line requires. But he is no longer strong. Nancy's therapist has pointed out that the caregiving will probably soon need to change hands. Nancy understands this inevitability, hears the sound of its approach in the ragged breathing of her husband asleep.

She is very much afraid. On very good days, she can imagine managing all of it — the whole house, the retirement, the illness, the love with pending losses. On bad days, the fear comes back, fear she has known since childhood, suggesting the worst is about to happen and proving that it sometimes does, fear that prevents her from leaving the house and sets up a cycle of catastrophic thinking that she paces out across the living room floor. Nancy still dreads nightfall and the insomnia that goes with it, knowing that insomnia is the trigger for manic attacks that are even more dangerous to her husband and herself than doom thinking.

Nancy is smart, practical, and aware of her own frailty. She needs to be able to leave the house, and she needs to be able to be at peace when she stays. Her husband deserves every care she can give him, and she, Nancy, deserves to believe that the end of his days will not have to be the end of hers too.

"Most therapists would have locked me up if I mentioned suicide as the only possible option," Nancy says, but when she was honest with her current psychiatrist, this one calmly suggested it was time to find the therapy that would teach Nancy that survival was possible . . . and desirable. A psychiatric service dog might be an option here, a dog taught to intervene and redirect Nancy's dangerous behaviors and comfort the night dread that keeps her awake. "A large dog and a calm one," the therapist recommended, "a dog with great physical presence, so you feel safe in its bigness."

Nancy read my first book, *Scent of the Missing,* and came to know my search partner Puzzle even better through the book than she had through our conversations online. When her therapist recommended Nancy find a service partner, Puzzle's nature attracted Nancy to golden retrievers. She says she's looking for a golden like Puzzle, with a kindly manner and a friendly expression, but she isn't sure where to turn or how to determine which dog is best. On a practical level, she is new to big breeds. Worse, Nancy's heart is wide open, and she knows she could choose a dog based on an impulse, a dog that wouldn't be right for the collaboration she needs to forge.

"Safe in its bigness." She repeats her therapist's advice to me, realizing that my Puzzle is petite and wondering how many goldens are larger. Having a big dog is quite a concept for someone with her particular memories, Nancy says. This will be a serious step, and despite every objection she has conjured (house size, fixed income, husband's health, situational cynophobia!), she has moved forward. And so the research and thoughtful quest for her partner has begun, with all its issues of age, health, and temperament required by nonprofit service-training facilities. Nancy describes her search as often discouraging, but she feels grounded and proactive. Is it possible that simply searching for the right canine partner is a therapeutic step?

One day, after weeks of e-mails, phone calls, and near misses, "The universe aligned and the gods laughed," Nancy tells me. Enter Lexie, a very light blond retriever from a bad situation who could use a little rescuing herself. On meeting, Lexie is friendly, calm, and gentle. She is also a big white dog.

The universe was looking out for her, Nancy believes, and looking out for Lexie, too, on the day that they first met. Taking her therapist's advice and the general good PR surrounding golden retrievers, a dog lover who found it easy to believe in their innate goodness, Nancy had launched herself hopefully toward a therapy she believed could work. She recognizes now that she had no clear picture of the training involved and the rare kind of dog that might be able to meet the demands of the work. Though big dogs were still frightening to her, Nancy had agreed to her therapist's recommendation. She agreed to a meeting with the breeders who were surrendering Lexie.

The reason for the surrender would have washed Lexie out of a formal service dog program. She had pupped, and pupped, and pupped again — litter after litter of beautiful light golden retrievers — and multiple pregnancies had been very hard on her hips. Lexie now had hip dysplasia, a condition that rendered her useless to her previous owners. She wasn't an old dog, but multiple, heedless breedings had aged her.

There is risk for the person who makes a choice of service partner that's driven by compassion, and Nancy says that even as she was moved by Lexie's plight, she was aware that she was choosing a dog whose limitations would also limit the ways the dog would be able to serve. Lexie would never be able to walk long distances on hard surfaces. For a younger handler needing daily assistance at home, at work, and in between, Lexie would not have been an appropriate candidate. But Nancy was already aware of the tasks she'd need from a canine partner, and long-distance walking would not be one of them. Nancy needed a dog who would disrupt the loop of her obsessive anxiety, a dog who would wake her from nightmares, and a dog whose presence would encourage her out of the house and into the neighborhood. Without help, without formal evaluation of any kind, Nancy was making the choice on instinct.

She was aware that the partnership might not work out. She also knew that Lexie needed an advocate, fast, and if Lexie couldn't ultimately be trained to assist her, Nancy could at least keep her safe and find her an appropriate home. Nancy adopted Lexie out of instinct, compassion, and what she says was a nudge she couldn't deny. Something good would come of it for both of them. She felt sure of that, even as she watched, with trepidation, Lexie enter the house. Nancy found that every movement from the dog was unfamiliar and a little threatening. The dog wasn't hyper. She wasn't aggressive. She wasn't loud. Lexie was just so big.

It took a month, Nancy says, before she found herself able to relax and really trust the dog that had given her no reason not to trust her. It helped that Lexie was gentle with Nancy's Pomeranian, Jerry, who was by nature timid with big dogs. It helped that Lexie almost immediately connected to Nancy, choosing to move with her through the house, be in the same room, and settle nearby. And it helped most, perhaps, that Lexie seemed to naturally pay attention to Nancy in a considered way. After her life as a kennel dog, perhaps she was as surprised by the gift of fellowship as Nancy was. The freedom of movement and the opportunity for companionship may have been as important to her too. Whatever the reason, Lexie had the gift of steady gaze, and she would watch her new owner with a soft eye and friendly interest. Nancy says it didn't take long for Lexie to recognize her partner's state of being, and after her first experiences with Nancy in periods of deep, cyclical anxiety, it took very little for Lexie to choose to intervene.

Nancy describes the wheel of her obsessive thought as being as unpredictable as it is maddening. The product of anxiety that can attach itself to anything ("even this phone call," she tells me during an interview), whatever it is that worries her presents itself, sparks an anxiety, is addressed, brings up a tangential problem, causes more worry, is addressed again — perhaps resolved, even — and then, in the brief, light free-fall moment after resolution, it introduces itself again. The pattern is so entrenched and the worry so compelling that Nancy finds it difficult, and sometimes impossible, to lift herself free.

What does Lexie recognize? Nancy says there's probably some change of posture or movement, a shift of expression that she cannot identify herself; she's pretty sure that there's a scent that she generates that's specific to all that anxiety. It's eerie, she says, to have a dog sitting calmly beside you who, within moments of the downward spiral of your own thoughts, decides to get involved.

Lexie's intervention is specific and persistent. She presses her head to Nancy's leg firmly, gazing at her, with a petition for petting that will not be satisfied by idle touch. Forcing Nancy to move, to engage, to pay attention to Lexie, the dog insists on a response. Nancy says that most of the time it works. She swims up from her dark place, buries her hands in Lexie's fur, and forces herself to concentrate on the calm, steady presence of the dog. The touch and engagement is often enough to break the obsessive cycle. Nancy says not always, of course. Sometimes she's just in far too deep, but she knows Lexie's determined interaction has saved her from real despair more than once.

How is this different from the presence of any loved dog? Nancy has had small dogs all her life. She says that in Lexie, there is a level of commitment that she's never experienced before. It can't be easy for a loving dog to be in the presence of severe depression and the desperate churn of anxiety, but instead of removing herself from Nancy, waiting for a better mood, or panicking in any way, Lexie calmly chooses to be close to her through the worst of it.

I mention another handler's comment: "With a psychiatric service dog, it's not a case of having to chase some pet dog down with the desperate plea: *I'm bad off. Let me pet you, damn it!* The dogs who really do the job know when to come to you, and they come to you, and they stay."

Nancy says, "That's it. They *stay.*"

Assistance dog partners, service dog professionals, the Department of Justice, and the public freely debate what does and does not constitute a service dog, and the argument is not always a clear one. As the ADA defines them, service dogs are trained to perform specific tasks designed to assist or mitigate conditions of physical, mental, or psychological disability. Much rests on the word *tasks,* and some of the most heated discussion centers on what defines a task and what does not.

A service dog that retrieves dropped items for her wheelchair-bound partner is performing a task.

A service dog that guides his blind partner across a street is performing a task.

A service dog that indicates to his deaf partner that someone is talking is performing a task.

A service dog that reminds his partner to take medication is performing a task, according to Joan Froling of Sterling Service Dogs and Dr. Joan Esnayra, of the Psychiatric Service Dog Society. Service Dog Central, "a community of service dog partners and their trainers," disagrees. From their website: "A dog trained to remind a handler to take medication, though helpful, would not truly be needed if the person was able to remind themselves to take their medication in ordinary ways, such as using an alarm." This might seem a small point to many, but there is any amount of debate over the person who is able to pay attention to an alarm and the person who is not and how, for one who is not, a service dog makes a substantial difference.

Such debate extends to the finer points of psychiatric service dogs versus emotional support animals. As defined by the Department of Justice, a service dog serves someone who has a defined (read: "authenticated") disability. The dog's tasks must: (1) be trained and not natural behavior, (2) mitigate the disability, and (3) be specific to the handler's needs. By this definition, a dog that provides comfort or a sense of safety and eases despair simply by being present is not performing a task. While the Department of Justice and others note the benefits of dogs in such situations, these dogs would be categorized as emotional support animals rather than service dogs because they are not performing a *trained* task.

One trainer tells me, "You can get a little lost in all that language, because honestly, one of the things we try to do is shape a stable dog's natural behaviors to the task we need them to perform. So in the case of someone caught in an obsessive loop, a dog that comes to his handler and insists on interrupting that loop is performing a task in a way that a dog that simply comes when called is not. (Even though, as any dog owner knows, a dog reliably coming when called is often something you really have to *teach* them to do.) In the case of the psychiatric service dog, that dog may recognize symptoms without command, that dog initiates the re-direction, and the dog stays in order to fulfill the task. But think about it — to an outsider, the dog performing her task may look a LOT like some dog just asking to be petted."

Nancy doesn't worry about such distinctions too much at the moment. She's taking this therapeutic approach slowly, small steps outbound. She is severely agoraphobic and dislikes leaving the house at all. At this point, Lexie's service to her is most crucial at home. Nancy sees a change in herself since Lexie's arrival. She cannot deny the good work of this dog who can disrupt her obsessed periods, wake her from nightmares, and stir in Nancy a willingness to go outdoors and engage with her neighborhood. She goes out voluntarily and she's learning to enjoy it, but she primarily goes out simply because Nancy feels she owes it to her dog. It's a progress to be celebrated. For the first time in years, Nancy advances rather than retreats.

It's a change that eventually brings its problems too. Though most of her neighbors are encouraging, one objects strenuously to Lexie's presence. Their gated community is designed for residents over fifty-five, and policy changes now include a size limit for incoming dogs. That policy never worried Nancy before. Her pet dogs were always small. But Lexie is a new dog, and under the new rules, she exceeds the size limitation. Nancy has brought Lexie in with all the credentials required of an ESA or a service dog in training and has shown them to the administrators of the homeowners' association, but the objecting neighbor doesn't believe a word of it. Nancy has no disability that the neighbor can see. The dog is "a mutt masquerading as a service dog" and Rules are Rules. She directly, repeatedly confronts Nancy in front of other neighbors.

The attack could not have come at a worse time. Nancy has been struggling with deep depression for weeks, a depression exacerbated by the shooting attack on Gabrielle Giffords that claimed six lives and injured fourteen others. The violent event is inescapable here in Arizona. Local and national media are saturated with the story, which streams heavily across social media as well, and Nancy, sensitive to violence, whose compassion and empathy are on overdrive and who is already inclined to withdraw, has retreated even farther into herself. Lexie seems to recognize the distress. She steadily watches, refuses to leave Nancy's side, and travels from room to room with her through the house, even when Nancy sometimes sits in one place for hours, spiraling downward.

Leaving the house for any reason is a struggle, but it is important to Nancy that her dog has a healthy life too. Lexie's needs draw her up from her own condition. She takes Lexie to the community dog park — a simple, public act for most dog owners, but in this moment terribly difficult. "You have no idea how hard depression with social disorders can be," Nancy tells me. "When it's bad, you'd sooner walk on glass than face strangers, even acquaintances." Nancy forces herself anyway. Lexie's playtime at the dog park is important to them both. Lexie gets to roll in the grass and chase a ball like any other dog, play like a puppy with the others, and, most important to Nancy, Lexie doesn't have to focus on her for that short time. "We all need downtime," Nancy says, "and I think [psych service dogs] often don't get much of it because of the nature of their work. Mental illness doesn't take time off." Nancy is exhausted with her own despair and wants Lexie to have some relief from it.

But the vocal neighbor's reach extends even here, where her verbal bullying and derision have worsened Nancy's already fragile condition. The dog park has been the scene of some of the worst of it, and Lexie now refuses to play or leave her partner's side. Nancy is moved by her dog's dedication, and at the same time, she's also furious the situation has escalated to this. She hates confrontation, but anger and concern for Lexie's well-being prompts her to respond to the verbal attacks in kind. She reports the neighbor to management even as the neighbor complains. When management responds to the neighbor and tells her that Lexie is in service to Nancy and has full permission to live there, the neighbor presses for information about Nancy's condition, which the managers refuse to give. The neighbor is told to stop the harass-ment, that it could have legal consequences — a definitive ending to a situation that has gone on too long. Nancy is proud of herself for re-fusing to hide and grateful to be beside a dog like Lexie whose service presence has changed her and whose own needs are worth protecting.

It's over, but it's not. Sometimes the harassment still resonates. Nancy says that sometimes looking "normal" is the hardest part. There is nothing unusual about her appearance. She knows it's difficult for others to understand Lexie's presence or to recognize how vital a role she plays in Nancy's ability to even leave the house. She understood there would be stares and questions about Lexie, and on good days she thinks of calm, prepared answers. Some days, riddled with the relent-less nature of her condition, she cannot face going out. Sometimes a single question is just too much. Lexie's assistance, her presence be-side Nancy, is undeniably worth it, but the weight of public scrutiny is sometimes more than she can bear.

What is it that gives some dogs such a depth of compassion for humans that other dogs — sweet, amiable, friendly as they might be — don't have? Months later, after the trouble with the neighbor has faded, the community dog park provides Nancy another opportunity to understand Lexie. Normally divided into two areas, one fenced side of the park has been closed for maintenance, and now the tiny dogs and the larger or more aggressive ones must share a single space for the duration. It's an uneasy peace. Many of the residents with smaller dogs are fearful for their pets' safety, which Nancy understands, and Lexie's size worries a few who don't know her. Nancy understands that too. So the dog-park mingling isn't as easy as it might be. Conversations are awkward, a little wall-eyed, with the supervision of unfamiliar dogs as playmates.

Lexie is enjoying herself among the others when Nancy sees her dog suddenly stop playing and cross the park to approach a seated woman. Pressing herself near, Lexie licks the woman's hand and places her head in her lap, and even from a distance, Nancy can tell that the woman is upset. Unsure if she is frightened by Lexie's size or something else, Nancy calls to Lexie (who either doesn't hear her or, Nancy admits, ignores her). When the woman begins to weep uncontrollably, Nancy, moving quickly, calls her dog again. What is going on? Nancy isn't sure, but she's worried that somehow her dog has upset someone. By now other bystanders have approached, and several women put their arms around the distraught figure, Lexie still pressed to her side. To Nancy's surprise, the woman bends down and buries her face in the dog's fur. Nancy stops where she stands, paused by that need. She learns from a neighbor that the woman's husband had died the day before. The woman and the golden retriever remain in that position for a long interval. Nancy doesn't interfere. That her dog responded to the woman's heartbreak doesn't surprise Nancy, but how did Lexie know? "Is this scent or sight or something else?" Nancy asks me. "How does a dog recognize grief in a stranger half a dog park away?"

# 22

EVEN AMONG MY PACK of rogues and scoundrels, Mizzen is an oddity. The dogs aren't quite sure what to make of her skittering gait, her constant chortle, and her strange maneuvers that lack any kind of dog logic. She stares at walls, sometimes for a half an hour or more. She likes to rest with her backside propped up on a plant stand and her forepaws on the floor. I once found her asleep in the water dish. (Water was still in it.) And of course, she likes to feed her head to Jake.

Jake Piper is patient with Mizzen, whose devotion sometimes makes it difficult for him to simply walk through the house. He steps over or around her; he tolerates her constant weaving back and forth under his belly as he bends over the water bowl. He is as patient as she is extraordinary, washing her face now and then, occasionally opening his mouth to let her stick in her head and grumble.

The other dogs are less tolerant. Puzzle watches Mizzen from a bemused distance. She isn't hostile, but the chocolate Pom, with her pointless *Hoor! Hoor!* explosions (even Sprits'l doesn't seem quite sure what she's barking at), her bouts of spinning, and pig-at-a-trough behavior at the water bowl doesn't appeal to Puzzle at all. The Poms don't know what to make of Mizzen's oddness. She's their size, their shape — she may even share a common Pomeranian scent with them — but as weeks and months pass, they never warm to her. They stand away and watch the event of her rather than engage. Mizzen doesn't seem to care. She is very fond of humans; she gives and receives affection well enough, but Jake is the center of her universe.

When work in the backyard brings contractors to the house, for a time I have to walk all the dogs in the front yard. Jake Piper and Puzzle come out with me off-lead for a game, but the little Poms have to be walked singly or in pairs and on leads. The gaps between the pickets on our wide fence are enough for a Pom to squeeze through, and we live on a fairly busy street. So out we go, the small dogs on the end of longish leads, enjoying their constitutional.

None of the little guys enjoys walking with Mizzen. She behaves well on a lead, but when they're connected to her by virtue of me, their wariness increases. Though it's not outright hostility, I can see their whale-eyed side glances. I try to imagine how they frame her: Is it the pale eyes; the random, spasmodic gait; the chatter over nothing they can see and hear? I don't know. But Mizzen is the first dog that's ever come into the house and remained such a subject of speculation.

It's difficult for me to watch. Full of sympathy for what I imagine is her feeling of rejection, I begin taking her out alone. This seems fine with her, just as walking with the other dogs is fine with her. Everything is fine with her (*hor-hor-hor*). She's her own strange universe, is Mizzen, and while it would be easy to frame the odd behavior as diminished intelligence, I can't. I'm too curious. I privately begin teaching her the Door command from outside, and we work on other small tricks she seems delighted to do. Apart from the other dogs, she blooms in small ways. She seems a little more sensible and a lot less strange.

Sometimes we have a human audience. I live on a corner. When Mizzen is out in the yard with me, her size, gait, and color provoke second glances from passersby. Is that a squirrel? A rabbit? Or some other kind of wild animal? She doesn't read *dog* to many people. There are plenty of joggers in the neighborhood, lots of pet owners walking their dogs. Some of the dogs strain to meet Mizzen with friendliness, and some look like they could have her for lunch. There are delivery persons and contractors rattling down the street in big vans or in trucks full of lawn mowers or ladders. The occasional kid wearing earphones swifts by on a skateboard, cap slewed sideways, bored expression in place, weaving down the middle of the street in a roar.

From some people we get waves and comments, which Mizzen doesn't seem to notice. If Mizzen is engaged with you, she pays attention to you. She doesn't flinch at the clatter of trucks or skateboards; she doesn't remark at all on the yip and skip of passing dogs; and, again, if I didn't see her ears flick in acknowledgment, I might think she was deaf. We linger on the wide lawn, and I tell Mizzen that she's got focus. She is special.

Some days we are visited by a lady about my age pushing her mother in a wheelchair. They are both slight figures, variations on a theme, the elder woman papery with age; her white hair cut short is thin as the feathers of a baby bird, revealing the curve of her head. In the long shafts of afternoon sunlight, she glows like a dandelion. She always wears a bright housedress. A clutch of stuffed animals and a baby doll travel in the chair with her.

Even though her mother is a small woman, the daughter struggles a bit with the wheelchair. It is old. One of the handgrips is wrapped in duct tape. The daughter sometimes has to lift up the back of the chair and give it a little shake, as if a wheel is misaligned and sticking. Mizzen and I often hear the rasp of the chair before we see the two women, and lately we've begun to stay outside if we know they are coming, because mother and daughter both love Mizzen. The first time they saw her, they slowed as they progressed down the side of the yard, and then, after turning around at the end of the street, they came back and stopped to watch Mizzen learning her tricks. They didn't seem to speak much English, and my Spanish isn't strong either, but I caught the word *osito* and then *perro,* and I knew they were talking about Mizzen.

"Would you like to pet her?" I pointed to the dog and pointed to them and made a little petting gesture, and they laughed — the daughter first, the mother a moment behind — the same laugh, a flash of twin expressions. The daughter shook her head politely while her mother stared over the fence at Mizzen. But they made no move to leave. They couldn't take their eyes off the little Pom, and when I led her to the fence, the daughter came over and bent to touch her, then extended her mother's hand and wiggled her fingers, as close as she could get her to Mizzen from where she sat.

The mother deserved a chance to pet her, and Mizzen deserved the chance to be openly adored. I carried the little dog through the gate and around to the pair and knelt before the wheelchair, giving them access to her ears and her soft fur. Mizzen was oddly careful in their presence. She lifted her nose and squinched shut her eyes, the stub of new growth on her fluffy tail wagging the way a puppy's does, a furious joy that tickle-brushed across my face and made all of us laugh.

That was the beginning of the visits that continue still, though they happen at random. Some days we might see the pair twice a day. There have occasionally been weeks when we don't see each other at all, a condition that used to make my heart catch, concerned that something might have happened to mother, or daughter, or wheelchair. Soon it made better sense: their presence or absence is driven mostly by weather. Neither mother nor daughter can bear the heat. A strong wind makes the wheelchair harder to push.

As we meet across months, our language skills don't improve much, but we have a few rituals we've come to understand. Mizzen moves from my arms into the daughter's, and then onto the mother's lap, where we both hold her until she settles. The elderly lady has had a series of strokes, it seems, and in the time I have come to know them, I've seen one hand fail, and parts of her face seem to go slack, giving her a harsh expression that seems cruel for such a gentle soul. There are other issues. Alzheimer's, perhaps, or another form of dementia — there is a childish vulnerability that is more pronounced some days than others. She speaks rarely. When she does, her voice slurs across the Spanish, making it even more musical to my ears. Though her daughter always bends forward to hear her, their communication seems almost free of speech. The daughter cradles her mother's shoulders when the wheelchair isn't moving, sometimes folding over the wheelchair to press her cheek to the soft down of her head, guiding her hands to curve around Mizzen's back and holding them there. The moment is bittersweet. Sometimes the daughter turns her face into her mother's temple and closes her eyes.

Mizzen sits calmly with them. In the house she can be a little hyper, and I wasn't sure how well she'd adjust to the senior lady's frailty, but somehow she's able to contain herself and lie quietly, enjoying the tentative strokes on their own terms. She seems to like it best when both mother and daughter are touching her. She relaxes into the caresses, burying her nose into the cupped palms, then looking up into their faces. Sometimes, in the way of dogs being adored, she will look at me as if to make sure I am noticing it all. She croons a little, Winston Churchill rumbling low, when her ears are rubbed. She enjoys a gentle inspection of her white-gloved paws. There is not another dog in my house more suited to these visits than Mizzen.

I am grateful. If I spoke better Spanish, I could better express to mother and daughter what their affection brings to Mizzen too. They've made her special, something I think she can sense. If I spoke better Spanish, maybe I'd give them my phone number or tell them to knock and ask for Mizzen if we aren't outside. But if I spoke better Spanish, or they spoke better English, perhaps we'd make appointments we couldn't keep, obligations we'd feel guilty about failing later.

*Buenos días,* we say for hello and goodbye; nods and *gracias, gracias* all around.

Once, after they set off again, the daughter turns from her grip on the wheelchair. She smiles at Mizzen first and then at me. She says in English: "Every day we come for the little brown bear, we are good." Then she turns back to the wheelchair as the wind catches both of them — a harder progress, a rasp of wheels, her mother's fine hair lifting light and free.

# 23

I am standing in the kitchen, ringed by curious dogs. They are watching me sneeze for the sixth time, perhaps waiting for some kind of levitation. A series of sneezes in rapid succession, separated only by pauses while I stare stupidly, fixedly, open-mouthed, waiting for the next one that's already burning in the chute to actually happen. The dogs have heard me sneeze before, but really, this is something extraordinary. Kuh-*chow!* Kuh-*chow!* Kuh-*chow!* Kuh-*chow-chow!* Kuh-*chow!*

Just before another one, Jake Piper comes forward to sit by me, rests one paw on my right foot. When I sneeze, he leans away a little, ears swinging back.

It's my own fault. I've got chalk on my hands, and I'm allergic to chalk. Now I've absently rubbed my nose. I discovered the chalk allergy thirty years ago when I was a student teacher in the days before whiteboards, but I had forgotten all about it when I got the bright idea of a little chalkboard in the kitchen to post lesson plans for each dog as we train forward. Just the kind of absent-minded, kitschy little idea I come up with, and now, as I squeak across the chalkboard, making the dogs' heads tilt, reminding myself to wash my hands, I think that at least I've amused them. They all came into the kitchen to see what was up. They all remained for the show.

"Jake," I say, "these are your marching orders, buddy."

I've lost some of my long-ago chalkboard skills and have erased several times to make the words legible, then to make them neat across the slate.

The chalkboard reads:

## JAKE
### TASK: REDiRECT REPETITIVE BEHAVIOR

The writing is crooked and unreadable. I erase and rewrite it again.

With Merion and others with similar conditions in mind, I am about to begin training Jake to intervene in sample OCD behaviors. There is a wide range to choose from, some easier for a dog to redirect than others. OCD refocus is an important skill for psych dogs partnering handlers with the condition.

While OCD was once considered rare, which might have been due in part to limited understanding of the condition and patient reluctance to disclose behaviors they recognized as irrational, that is no longer the case. The National Institute of Mental Health reports that about 1 percent of the adult American population suffer from the disorder —about 2.2 million people—and approximately 0.5 percent of that group are classified as having severe symptoms.

A search-and-rescue colleague once remarked that she has known plenty of OCD behaviors in first responders. "Firefighters check stoves," she said. "Police officers get hung up on locks." That was an uneasy comfort to me in 2004, the awareness that maybe I wasn't alone in my reaction to the search field, the recognition that for many people, the work can strike very deep.

People with the full-blown condition and mental health professionals alike describe a life with OCD as a life in secret chains. Whether fearing germs and compulsively washing, frightened of harm and compulsively locking, holding rituals or numbers as a talisman against catastrophe, or hoarding goods, even trash, out of fear of future need, those with OCD have a deep anxiety to soothe at the same time they often recognize the unreasonableness of their actions. Many describe being terribly embarrassed. They withdraw socially and avoid mentioning repetitive behaviors to family and therapists.

Dr. Steven Phillipson and others suggest that many people who don't have clinical OCD have nonetheless experienced repetitive thoughts (a nagging, perhaps overblown, worry that lasts for days; a song you can't get out of your head) and performed acts not far from the OCD array (always using a foot to flush the toilet, habitually knocking on wood to prevent a bad event from happening). The line can be a fine one. What constitutes true OCD? Phillipson suggests that one simplified test is to ask yourself what amount of money it would take for you to *not* perform the habit that results from the troubling thoughts. Those without OCD might cheerfully stop the behavior for ten to a hundred dollars. Those with OCD might find it impossible to give up even for a hundred thousand dollars.

While I have known this ugly cycle, and it took only an insistent puppy to lead me free of those habits (I might have said, *Yeah, I can give this up for a hundred dollars,* and maybe I could have), it is not difficult for me to imagine how overwhelming the condition could be at its worst. Respondents on mental-health forums post about it with equal parts fear and frustration, and behind their online avatars, they describe their own struggles.

There are those with an obsessive desire for order, placement, or symmetry, which they believe keeps some bad thing—joblessness or poor health or the death of a family member—at bay. There are lucky numbers that must be repeated in order for some to survive the day, objects that must be vanquished by some ritual behavior or saying. There are those who never touch a doorknob or who scrub their hands raw after picking up the mail—in gloves. Some talk about sleepless nights caused by fear of the stove being on. They go to bed, worry about the stove, get up, check the stove, go to bed, wonder if they really checked as carefully as they should have, get up, check the stove, go back to bed, get up, and so on. Hours of checking and rechecking, lost sleep, anxiety soothed only at the moment of touching the stove and then renewed the moment that contact is lost. And then, one woman writes, there is the anxiety produced by the fear that all of this means she is completely losing her mind.

One young man I talk with calls his OCD "the boa in the basement" —this huge *thing* not visible to outsiders, always there, mouth open wide, ready to consume him if he doesn't feed it. Light switches are his nemesis. He feeds his fear by flipping them. A whole house of light switches: if they are off, they are turned on and then off again; if they are on, they are turned off and then on again. What is he afraid of?

He can't really say. Electrical fire, maybe, if the switch is on. Darkness, certainly, if the switch is off. The flip of a switch gives him a brief reprieve, and then the snake's mouth opens once more. The boa is a perfect metaphor by a guy who'd like to write a book, he tells me, a work-from-home guy by trade who lives in a two-story house and finds it increasingly difficult to get his job done because the boa gets bigger and more demanding by the year. He estimates that on a bad day, he loses three to four hours flipping switches through the house.

Psych service dogs, in conjunction with cognitive-behavioral therapies, can be a big help to partners with OCD. Fully trained to redirect the handler's compulsive habit, the dog alerts the person to the behavior, then offers an alternative one. Some of these behaviors are typically provoked by departure or bedtime rituals, so the dog might urge the person toward either the door or the bed, offering affection or play as a distraction. Some, like Merion with her assistance dog Annie, train their dogs to bring the leash, "suggesting" a walk. The service dog acts as a reality check, a refocus, and a healthier reward for the handler.

Nancy writes:

> For most people, there is extreme embarrassment involved [with OCD], because to the observer, it makes no sense whatsoever. . . . It is an extremely hard cycle to break because of the nature of anxiety → reward → anxiety → reward. Reward being a break in the anxiety for a short time. For me, to interrupt the cycle, it needs to be a larger reward than continuing the behavior. Enter petting/cuddling silky soft Lexie who deposits herself in front of, on top of, in the way of me. I cannot resist her love, and closeness with her has an instant calming effect on me. i.e.: reward. Does it always work? No, but it works enough of the time to be hugely beneficial to my life.

For Jake's refocus task, I choose a common compulsive habit. It is one that I have never had but understand: double-checking that the stove is off.

That Jake naturally wants to be where I am is a bonus. That the stove is in the *kitchen* is another one. All the dogs tend to keep an eye on the kitchen, so getting Jake to follow my movements there is probably not going to be too difficult. Even though stove-checking is not my problem, Jake and I need to live as though it is. The first step is for Jake to attend to my movements at the stove. On lesson one, I simply want him to stay near me and watch me work there. I call him to sit, I tell him to stay, and I begin cleaning the stove. The Sit and Stay commands he is used to, but he's a little bewildered by the commands in this context. He holds his Sit and Stay politely for a few minutes, but out of the corner of my eye, I can see his ears swing back and his posture shift. He's had about enough of this. He doesn't get the point.

"Hold your Stay, Jakey," I tell him, and when he is back in his attention posture, I give him a treat. Clean a little more. Give him another treat when we release. Lesson one over. The day outside is sloppy, but we go out to play ball anyway. I want to reward the good-natured, willing dog that seems happy to learn just about anything, even if it makes no sense to him. I watch him shake loose all the energy that being still and paying attention requires; he reminds me of students I've seen dance their way down the stairs after a final exam. The first time I throw the ball, Jake is so ready to play that he almost explodes, jumping so high and awkwardly I'm afraid he'll take a hard fall. We play hard until he flops down in the grass and pants a smile up at me.

An hour or so later, we try the same lesson again. When I move to the stove, Jake needs to be there, out of the way but watching. We will try this over and over, across days, with variations of time and activity. Jake will watch me clean and cook — in fact, I do more cleaning and cooking in this period than I have perhaps ever done, a bonus the first time I take my lunch and enjoy roasted vegetables in a soup broth I've made, with Jake as my audience. As long as I'm working at the stove, all he has to do is attend.

But now we need to train to trouble. There's no shortage of posts about stove-checking behaviors on the Internet, and I can sympathize with almost all of the descriptions. They sound much like my own intermittent problems with locked doors. In fact, some people posting on message boards talk about their anxiety as a beast in motion: they have transitioned from checking doors to checking the stove or from the stove to the dryer (one even from the dryer to the curling iron, double-checking that it was off, unplugged, and in the drawer), then back to the stove again. What is the meaning of all that? they wonder. Many respondents describe simply standing at the stove and staring at the knobs, looking hard across each knob to make sure the indicator is turned to Off. Some have to touch the knobs and attempt to turn them, feeling for the pressure of the Off position. Some touch the knobs and say "Off" aloud.

What to teach Jake Piper? I decide to have him intervene if the handler touches knobs repetitively, says "Off, off, off, off" in a sequence, or simply stands and stares at the stove. Getting Jake to interrupt me touching the knobs or repeating the word *off* seems easy to train. The clear behavior or the significant word becomes a cue for action. It may be harder to train Jake to interrupt a frozen stare.

And what should be his action? A few handlers tell me of their missteps in this direction. Some encouraged their dogs to bark then realized how annoyed they were getting with the barking that they themselves had asked for. They came away from the stove more irritable than redirected. One taught her Lab to squeeze himself between the stove and her, but instead of being redirected, she began to accommodate his position, determined to touch the knobs of the stove from three feet away, bending forward awkwardly over the dog. One handler had the bright idea of having the dog raise up with his forepaws on the counter beside the stove — an in-your-face *Hey there* that she couldn't possibly ignore. She admits she (and her trainer) came up with that without thinking of her dog's history. As a youngster, he'd been a terrible counter-surfer, had managed to eat whole bags of cookies and once an entire green bean casserole without ever dragging anything to the floor. The handler remembered only that the dog was tall enough to stretch up to the counter, didn't consider that maybe he shouldn't. Intelligent disobedience carried the day. She thinks her dog knew better. The first time she and the trainer attempted to teach him the paws-up-on-the-counter response to her knob fiddling, the dog looked at her in surprise and some suspicion, as though he remembered stern correction for counter-surfing in his youth. Was this some kind of trick? He refused to do it.

It took a little while for the handler to recognize the conflict. Then she felt guilty. Worse, she realized that if her dog's paws were in close proximity to the stove, he could possibly turn the knobs. She ultimately taught him to nose-bump her knee and stay there, leaning his weight against her until she bent down to pet him.

Unlike the Poms, Jake Piper does not bark simply for the sake of it. He will flush a squirrel with a bark; he'll bark to let me know about strangers coming too close to the house. He'll bark at dog frenemies. But he's happy to communicate with his voice, and he's already learned the Speak and Whisper commands: the former produces a single *ARF*, the latter little more than a *wuf.*

We decide to go with a Whisper and a lean-against-the-legs alert, followed by a rope tug as a refocus for the partner and a reward for the dog. Jake loves tug, and it's a game the handler can't play halfheartedly — especially with a strong, bull-necked dog like Jake. As we train, I carry a rope tug in my pocket.

For the next step, I move to the stove, and Jake settles into his watch position. I begin by touching each stove knob and saying "Off" with every touch. The first time we do this, Jake is puzzled. I don't normally talk to myself, and in our quiet house, he's not used to free-floating words not directed to another human or another dog. In my peripheral vision, I watch him watch me, the head tilting, the wayward ears flicking. I touch, say "Off," touch, say "Off," touch, say "Off," and at the end of the sequence, I turn to Jake with a happy "Whisper!" and swing the tug rope out of my pocket.

Wow, Jake did not see that coming, and he porpoises up with a bark that makes the stove burners ring, grabs the tug toy out of my hand, and races down the corridor. Certainly not a Whisper, not exactly in control, but it's a start.

"My service dog helps me grow at my own rate, on my own terms," a handler tells me. As well-meaning as her family is, and as supportive, she feels she couldn't achieve the kind of growth she hopes for with only human help. For every week of good days, she has two bad ones, she estimates, and in the bad days, she can't help feeling she's failed her husband, "who tries very hard," and her children, "who have lost some of their own childhood to the problem of me." At night — and sometimes in daylight too — she becomes worried about windows, unable to shake the sense of voyeurs or intruders. She has been afraid of this since she was a child and watched old horror movies where something was always waiting just outside, but instead of fading as she matured, her anxiety deepened. She can't help checking that the windows are shut and locked, the curtains carefully drawn.

Before she got her service dog, she could ask her husband and children to help her stop checking the window latches and tugging closed the curtains, and she did ask them, and they did try—but sometimes she couldn't help doing it anyway. She knew it was an impossible situation. She asked her family to intervene and then argued with them about this thing she couldn't give up, and she always got the sense of their exasperation just beneath the surface. This might not be fair, she admits. She might have been projecting the exasperation she felt with herself. Every bad day, when securing the windows was an aching compulsion, she felt more of a failure in their eyes. On bad days, they lived in her created, perpetual gloom.

Asking her family for a kind intervention became an admission that she wasn't getting better.

And then came her smart, verbal husky, who loved her straightaway and was happy to learn his "tricks." He is happy to be with her and very glad to intercede, to block her from the window and "let her have it" in the long-winded, poetic garble that is typical of his breed. From her beautiful dog, she feels no judgment. She does this thing; he does his thing in response. The difference is, for the dog, it's no big deal. Beside him, she can make light of it a little. Her dog can often break the spell with his dog voice, and then they go for a walk. The walk is his reward and hers. Sometimes just the two of them head out, sometimes her husband and kids go too.

She says that once, not long ago, she had a bad round with the windows after supper, and her dog had to really work to get her free. She paid attention to her dog and turned away, and they went out for a walk that led them from dusk into darkness. She remembers her husky's steady *pad-pad-pad*, their determined walk away from the house and toward a sky that went orange, pink, and yellow, setting the leaves of trees aflame. Often nightfall brought her a sink of depression, but not this night. She and her dog returned home by the light of streetlamps, and as they approached the house, she could see some vague images through curtains that weren't quite closed. She could see the light and movement of her family, a trace of pattern thrown through lace curtains onto the grass of the lawn. She was surprised that the sight didn't scare her. She looked to her home as a stranger would. She remembers thinking that this looked like a house where happiness lived.

Jake has his own idea about intervention and rewards. We have progressed to the point now where he recognizes both the stove-knob touches and the words "Off, off, off" as calls to action. The moment I step into the kitchen, Jake is there, the other little dogs in tow. When I don't head for the treat drawer, the others turn away, but Jake remains. He is a patient boy that off-command still sits in the untidy way of a puppy, rolled back on his bottom, his long legs sticking out every which way. Out of the corner of my eye, I can see him watching me. Jake could never play poker. His ears give him away, and I can see his train of thought play across them — ears casually akimbo when I'm puttering idly, folded back like origami when things seem calm, even a little boring, and then perked the minute I approach the stove.

A Whisper and a lean into me was to be our alert of choice for the OCD intervention. Jake has embellished that alert to his own benefit. He watches as I approach the stove; he stands by as I cook or clean. He lets me have one, maybe two, repetitive movements there, and he intervenes. Jake has lately chosen a more dramatic alert, pushing against the side of my leg with a Whisper, then sliding down the length of my leg to fall belly-up on the floor. He wants his chest rubbed. Jake Piper is as charming as he is insistent. If I don't do it, he whispers again. And again. And again. It's a soft double syllable rendered from that upside-down position, and it's impossible to ignore.

A good friend is interested in all this, but she finds it curious that I've chosen to teach Jake Piper to intervene with stove compulsion, which is not my problem, and not with the door-lock checking, which sometimes is. Why wouldn't I teach the dog something I might need down the road? Isn't that a little evasive?

Her comments startle me. I never consciously decided against teaching Jake to intervene with door checking. I just didn't do it. When she challenges me to teach him, I agree. I agree quickly, as though to prove stove knobs were a casual choice, not some great form of denial, but

I'm pretty sure she isn't fooled. I'm not either, and I privately love her and damn her for making me realize how hard some of this process is. How hard it must be for many handlers out there who'd like to believe that this one part of themselves, this one thing, this problem they've had in the past has been overcome, and it's something they won't need a dog for. I've never assumed I wouldn't have a problem with doors again. I've just hoped very hard that I wouldn't. And now, when I think about all the training moments that will go into teaching Jake to intervene in locked-door checking, I feel a little sick. I haven't had this problem since Misty died. To teach Jake, I'll have to mimic it, and mimic it, and mimic it. Ugh. I think of the old superstition naming calls; people will not name the thing they're most afraid of, because the power of the name is enough to bring it.

The next day, Jake Piper and I start working on my problem with the door. I try to think of it in his terms:

*Here am I, dog, and here is she, my person, and we should be leaving, but we are not. We're nearly leaving, but now we're not. Oh — this is like the stove, only outside. It's like the stove, with the staring and the fiddling. It's like the stove, with a new repeated word.*

This is good work for a dog who loves being outside. He picks up the problem quickly, and in his happy presence, it's not the burden I thought it might be. Jake's enthusiasm is infectious. In his eyes, the bigger his intervention, the quicker we're away to some better adventure. Quick is good. Away is better. So when I double-check the door he leans against me, hard, and Whispers, Whispers, *Whispers*. I hear the teeth clicking. I look down at his bright-eyed *We've got this, let's go* face. I have to laugh when he starts grumbling because I'm too slow putting up my keys.

# 24

SOME RESCUERS CALL IT Racebook, the use of social media to save animals bound for euthanasia. What began as e-mail strings in the 1990s and crossed to AOL and Yahoo message boards shortly after has evolved into a 24-7 awareness effort that has famously saved dogs just minutes before scheduled injections. For those of us involved in animal rescue of any kind, every day of the week brings nonstop pleas across the Twitter feed or the Facebook news stream. Purebreds, mixed breeds, large and small. Dogs, cats, horses. Lizards, snakes, and turtles sometimes. Chickens. They are all urgents with sad backstories — they are in trouble, out of time. The pleas are as heartfelt as they are well intentioned.

Pleas can be disturbing too. I've been involved in rescue through the Internet for more than sixteen years, but I still haven't been able to harden myself to them, and on days when I am already overwhelmed with the saving — or the failed saving — of one dog, to face the unending stream of pleas on the social news feed every time I log on is almost more than I can bear. For every dog saved, so many are lost. Social media brings hope and breaks hearts with every post. The urgents are always with us. Compassion fatigue is real.

One Friday morning, my Facebook stream shows a clutch of desperate re-posts about the same little dog in California. His captioned photo, originally posted by the Angels for Animals Network, has 250 shares and 170 replies by the time I see it, on March 9, 2012.

*Huntington Beach Humane Society, Huntington Beach, CA. Small blind, deaf neutered terrier mix, gray, 15 yrs., "Todo" in poor health.*

*Imp#C000111. Owner admitted to Hoag on 2-7-12, and was later
discharged, only to come back and was discharged again on 2-15-12.
Records show she lives in HB. Was mailed a postcard on 2-26-12 to
come and get dog. Postcard came back today return to sender. We've
held dog way past legal requirement and it now appears it's been
abandoned. Shelter will only hold him until Saturday.*

The photo accompanying the post shows a dispirited, unkempt dark
gray mop of a dog huddled on a blanket, gazing sightlessly toward the
floor. He has the look so many shelter rescues have after neglect is followed by the isolated safety of a shelter. He is safe, sad, and disoriented
— alive in body, but diminished.

I reread the plea description. In terms of attraction factor for many
adopters, this little guy has everything against him. He's a dark-colored
senior, blind, deaf, and in poor health. A fellow rescuer says bluntly
that the photo isn't going to help him much: with that head down and
his awful, matted coat, he's the last dog on earth most families would
consider. He would be high on many shelters' unadoptable list, and it's
a wonder he has made it this long.

But some shelters will make special allowances for identified strays
whose owners are in duress; maybe that's been this one's good fortune.
Todo is likely still alive due only to the uncertain status of his ownership; the information in his write-up suggests some contact with the
owner had been made. After a little research, I can decode the local
references. Todo's owner was twice admitted to Hoag Hospital in New-
port Beach in February, and then — was phone contact made with the
owner? Did that contact suddenly cease? — a shelter notification postcard came back marked "return to sender."

Todo, likely a misspelled version of Dorothy's Toto (whom this little
dog resembles), is now not technically a stray. Some people following
the little dog's Facebook plea page are furious because Todo has been
abandoned. Some are more compassionate, speculating that his origi-
nal owner may have died. Perhaps a neighbor or a landlord marked the
postcard for return.

We can never really know, suggests one reasonable soul. Is it fair to
judge?

Hell yes, responds another. She says she's been too long in rescue. She's seen all the excuses, heard all the lies.

Whatever the history, Todo has ended up the way many animals do when their owners are sick and the pets are left with others who either cannot or don't want to care for them.

This is Friday. The shelter will hold him through Saturday, and after that the prospects look grim. Sundays are not a reprieve for many dogs in line for euthanasia. The Facebook posting group has put good intention into action. They have raised a certain amount in pledges to pay for Todo's release from the shelter, and Todo's plea page has as many voices offering money to rescue him as it has voices begging openly for a kind soul to give the dog a chance. The money is there! Just go get him! Someone! Anyone! But so far, no local has offered to step up for Todo.

What moves me so deeply on his behalf? There are dozens of pleas on Facebook and Twitter every day, worthy dogs in equal hardship and despair, and I offer help with such pleas where I can, but there is something about Todo I can't disregard. I feel the loom of his Sunday termination in the pit of my stomach. He has tomorrow left. He is three states and 1,463 miles away.

I learn quickly that the shelter noted in the Facebook plea is not the actual name of the facility holding Todo. He is at the Orange County Humane Society, which has a pretty website, a compassionate mission statement, and a phone number that rolls to an automated greeting. California is two hours behind Texas, but no matter when I call the Humane Society, the phone is never answered by a person. From hard experience, I know urgent online pleas for a dog can continue long after the status of a dog has changed, and before I make a long-distance effort on Todo's behalf — plane tickets, hotel reservations, rental car — I need to make sure he is still actually at the shelter.

The phone continues to ring and roll to an automated greeting, and I can't get angry. Every unanswered call suggests the staff members are all busy. I get it, but I keep calling until the last moment the shelter is open on Friday. No answer. I post my phone-contact problem on the Facebook plea thread, and while a couple of local Orange County people promise to try by phone or to physically go check on him if they are able to get free the next day, most of this Facebook rescue network is far-flung, and the local rescuers are already overextended.

I go to sleep on Friday not knowing if Todo is still at the shelter or if, by some late-breaking grace, someone has come forward to give the little dog a home. The dogs twitch around me when I wake up at 1:00 A.M., 2:00 A.M., 5:00 A.M., subtracting the difference between Texas time and California time and tallying how many hours Todo has left. On Saturday morning, I start calling again, and the shelter phone rings and rolls to automated messaging as before. A Facebook acquaintance reconfirms that she'll check on Todo if she can get free that morning. But she's not close by. It's a maybe, at best. I have exhausted all my contacts in California (many are away with their families on spring break), and now Todo has about seven hours left.

I've never used Twitter for rescue before. In fact, I struggle sometimes with followers who do nothing but tweet THIS DOG DIES TOMORROW pleas, because I know so many dogs do, and every plea hurts. But between calls to the shelter, on impulse I tweet in the peculiar idiom of 140 characters:

Susannah Charleson@S_Charleson
*On standby 2 fly to LA 2 rescue a blind/deaf terrier abandoned at shelter. Euth tomorrow. If they will release 2 me, he's got a home.*

It's not exactly a plea — just a fling of frustration and concern into the Twitterverse. And because 140 characters doesn't cover my night of worry, moments later I tweet that I still can't get anyone to answer the phone at the facility. Almost immediately I get a response:

Tricia Helfer@trutriciahelfer
*@S_Charleson I'm in Toronto this weekend but if u need help with the shelter rescue before u get there, let me know and I'll send a friend.*

And a moment after that, another:

Shauna Galligan@ShaunaGalligan
*@S_Charleson I can go get him if you can't make it!!!!!*

A plea asked and answered in less than five minutes. Actors Tricia Helfer and Shauna Galligan are not strangers to me, but knowing the odd hours and cross-country demands of their working lives, I would never have expected they'd be able to help.

Tricia, Shauna, and I haven't known one another long. We met in Hammond, Louisiana, not far from the set of the TV pilot adaptation of my first book, *Scent of the Missing*. Tricia was cast to play the role based on me, and stuntwoman Shauna was hired as her double.

TNT ordered the pilot, and when Tricia was cast in the role, in autumn of 2011, I was excited. She is a strong woman. I knew her work and had no trouble imagining her in muddy boots and SAR gear pressing through the wilderness behind a search dog. Tricia grew up on a farm, loves to hike, has a passion for motorcycles as well as some flight experience in small planes. A random Twitter photo shows her waist-deep in a hole she'd helped dig for a new fence on her family's farm. I had absolutely no voice in the casting decision, but I was pleased that producers had chosen someone completely plausible in the field. Her background and sensibility seemed right on to me — and when I learned Tricia is also heavily involved in animal rescue, I remember thinking that she would know what it is like to do something hard and compassionate for no other reason than it's the right thing to do. The day I left Dallas for the production site proved the point: Tricia was spending her day off from the shoot visiting an animal rescue near Baton Rouge.

Shauna and I also found a connection over homeless animals. We stood in the rain and mud one day while a new scene was setting up, and she described her own background with dogs in trouble. They seem to cross her path and she takes them in, knowing that somehow she'll be able to find each of them the right home. She could have ended up with a houseful of dogs she couldn't place, but she hasn't. It always seems to work out. Shauna's heart for these animals and her gut faith in the process is compelling.

After the TV shoot finished, we exchanged e-mail addresses and connected on Twitter. I returned to Texas, and Tricia and Shauna returned, briefly, to LA. The subsequent months took us all in different directions. Tricia traveled a lot. She was guest-starring on NBC's *The*

*Firm*, which was shooting in Toronto, as well as voicing video games and guest-starring in episodes of other series. During intermittent weekends back in LA, her homecoming was often marked on Twitter by splendid shots of her rescued cats in different attitudes: sly, athletic, conniving. Luxuriously asleep. One cat caught in slant-eared roguish charm followed by another sweetly shy.

Shauna was working as a stunt double across of a variety of projects, tweeting various gruesome shots of herself as a crime-scene victim or an unlucky passenger in an exploded car. There were happier photos: Between long days spent slashed and bloodied, Shauna posted pictures of her newest rescued dogs, an adult golden and a golden puppy who were now living a rich life beside her as they explored the good hikes of Hollywood's hills.

I was back in Texas, then went to Boston, D.C., Florida, and Arkansas for weekends — writing and working with the dogs whose pictures I also posted, not glamorous but appallingly honest: Puzzle elbow-deep in mud after a search; Jake Piper photographed in the Great Cheese Leave It, which was followed by his total destruction of a new rosebush in less than two minutes flat.

While Shauna's and Tricia's local paths crossed often, it's safe to say that my connection with them happened mostly on Twitter, life fragments in fractured text and a clutch of photographs invariably involving rescued dogs and cats looking happy and often more than a little smug.

And then came Todo — the blind, deaf, gray myth of a dog abandoned in a shelter not far from Hollywood. With one tweet, he would be saved by six people a whole country apart.

The Orange County Humane Society, a nonprofit organization, makes it clear on its website that adoption is a far more considered process than many people might think. The staff's interest is in the welfare of the animals placed in their care, many of whom come from extreme hardship, and also in making a match between animal and family that is best for all concerned. The shelter's procedure includes a pre-application screening, proof of identity, proof of ability to house a pet (a rental's pet policy or proof of home ownership), and a suggestion that the entire family, including other dogs, be brought to the facility prior to a pet choice being made. There's a $130 adoption fee at the end of the process. The adoption policy on the site is firm and unapologetic, but kind. The intent is not to frustrate would-be adopters but to protect the animals involved.

As a rescuer, I can only applaud these kinds of policies. Too many animals are adopted out of casual interest or what some rescuers call "cage compassion" — the good impulses of the moment during a store adoption event that don't survive to Monday morning, when the new animal chews a shoe or whimpers or has its own needs an hour before the adopter really wants to get up. There are worse outcomes too. Some shelters, particularly those with very low adoption fees, have also been hard hit by "adopters" fronting for dogfighting rings. These are professional scammers in every sense of the word. They know how to dress, they know what to say, they bring shining children as proof of goodwill. The end result is horrific.

Shauna Galligan and actor Mark Derwin, coordinating by phone with Tricia and her husband, Jonathan Marshall, made the trek to the Orange County shelter to do Todo's shelter pull for me. Though I kept trying that entire Saturday, I hadn't been able to make contact with the shelter, and because time appeared to be running out for the little dog, one of them would formally adopt him for me and care for him until I could get there. It would be a complicated process requiring a sync of schedules, a crosstown drive in LA traffic, and someone in this spontaneous rescue team of ours finding a way to provide proof of home ownership despite the fact that most documents were locked up in business managers' offices on a Saturday.

Somehow Mark, Shauna, Jonathan, and Tricia achieved all that in a ricochet of e-mails and phone calls across North America. From Toronto, Tricia e-mailed me updates of their paperwork gymnastics and their progress toward Todo.

Finally, she e-mailed:

*We got him!!! Shauna and Mark are taking him tonight. The shelter is shaving him now.*

*We got him! We got him!* I post on Facebook, where Todo had an anxious following, and in moments the rescuers and dog lovers worldwide who knew about him began replying. Between his plea entry on Facebook and the ongoing updates on my own page — my Facebook friends were involved in the story unfolding too — Todo was well known before he ever left the shelter. Someone who had seen him mentioned that when she was there on behalf of two other last-chance dogs, she'd inquired about Todo and learned that shelter staff had been making every effort to get him the attention he deserved. "He's such a cool little guy," a volunteer said, describing a sweet, engaged personality that was slowly giving way to despair. This was a dog that loved people. There had been no interest in him from anyone for way too long.

Shauna writes:

> Let's see . . . we didn't go past the front desk. We asked if we could adopt [Todo] and then they asked [if] we wanted to see him first and we said no we just want him. And quick! . . . They brought him out and he had very long fur, very dirty and stinky most of all. They asked if we wanted them to shave him and so they did. He still smelt so terrible and poor guy had big scabs all over his tiny body. He acted as though he had just given up on life and didn't care what happened to him . . .
>
> He had been in there for two months and they contacted his owners but they never came to get him.
>
> My heart went out to him. He was in awful condition.

Todo had arrived at the shelter in rough shape. In addition to having an ear infection, he was blind in both eyes. Shelter staff had been advised that the left one might need to be removed. He had been given the best care available on limited shelter resources.

Shauna forwards photos of the little dog, shaved to the skin, wrapped in a towel, and collapsed in the back seat of a car. He looks frail and exhausted. He is on his way to Tricia's house, where Jonathan (who is coming from three hundred miles away) will take him to the vet. A later photo from Shauna looks more hopeful. Todo is awake, upright, resting on his chest in deep grass. He seems to be enjoying a lovely California evening and the feel of spring sunshine on his skin. With that coat shaved down to nothing, he is a skinny, scrawny little thing with button eyes and a mouse nose. He is a cartoon rendering of a dog. His upright ears are impossibly big.

"Holy ears, Batman!" says a friend who sees the picture. "That's not a dog. That's a rabbit."

It is something very like a rabbit I see Jonathan Marshall holding in a photo taken at the vet's office later that evening. Jonathan looks triumphant. Todo looks better. He is resting comfortably in Jonathan's arms, his ears up and face alert, one forepaw casually extended. Perhaps it's just wishful thinking on my part, but the little dog seems brighter, as though somehow in his recent passage through kind hands, he has sensed that things are looking up.

Heart murmur, total blindness, uncertain hearing, arthritis.

Other than that, Todo is in "pretty good health," reports the vet who attends him. Jonathan sends a video of their gentle first interaction. I can see blind Todo's cautious way-making on frail legs, and the vet's soft examination of his body and his movement. He is at last getting the veterinary care he's needed for months.

We are trying to figure out how I should bring Todo home. The heart murmur, of uncertain grade, is a concern. His current level of exhaustion is too. A plane flight in the cabin at my feet would be a shorter but more intense process than a drive. Would a four-day car trip truly be less hard on him? It's difficult to know how to bring him to Texas. When Jonathan reports the next day that Todo isn't eating or drinking and has done nothing but sleep, I can't be sure it's best to bring him here at all. I put out the word with my California contacts that we may be looking for a local adopter; if someone falls hard for Todo, we can avoid the stress of a long-distance transport entirely.

He's something of a worry across the next several days. The little terrier-Chihuahua-papillon (possibly) mix refuses food and sleeps almost nonstop Monday too. Jonathan has installed him in his own home office, given him a little dog bed and a plush towel blanket, and, because the little guy seems to tremble all the time, made him a sweater from an athletic sock, cutting out holes for paws and head.

Todo also has a new name. Shauna and Jonathan agreed that he needed a name that reflected both his hardship and his survival. Todo was no longer appropriate. Oliver Twist was good; Ollie T, the twenty-first-century upgrade with some street cred, even better. On the second day after the rescue, Jonathan sends a photo of Ollie in his rad sock sweater stretched out on their lawn. He still isn't eating, but he's awake and alert. It is a bright shot of a blind dog enjoying sunshine and soft grass. Ollie's head is up; his forepaws are stretched forward luxuriously; and his perked ears and wide smile, mouth open in a pant, suggest a dog outrageously happy.

Now, if he would just eat.

Newly back in town from Toronto, Tricia takes him to the vet for a second checkup, an evaluation for flight, and any documents Ollie might need to travel. The second vet evaluation seems more hopeful. Ollie is responding well to the medication, is brighter with rest, and he is overall a stronger dog, even though he's still not much interested in food.

I make a late-late-night flight to Los Angeles from Dallas, and it is a bright, beautiful morning in LA when Tricia takes me home to meet Ollie. He's been camped out in Jonathan's office since he came out of the shelter four days before, visited and scrutinized by their rescue cats — particularly Cesar, an extremely social orange tabby who wants to make friends. Ollie is less withdrawn now and has begun to explore Jonathan's office a little more. Previously subdued and so silent he seemed to be mute as well as deaf, Ollie surprised Tricia earlier that morning with a bark. It was a real-deal attention-getter of a bark that she heard one floor down and three rooms away. He was awake, he wanted something, and he didn't mind barking to get it.

He is awake when I arrive. Tricia carries him out to meet me. She's experienced and gentle with rescues. He is smaller than I had imagined. The little dog is quiet as she holds him, again lying easily with the left foreleg extended, and I get the sense that in his previous life, Ollie was very much a dog in arms. I am used to holding senior dogs, but when we shift him from her arms to mine, his fragility startles me. I can feel the hammering pulse beneath his left foreleg, the ribs and hipbones jutting just beneath raw skin, and the trembling that everyone has noted since the day he came home. Whether from age, chill, or the trauma of change, Ollie shakes all the time. He trembles as I hold him, even as he leans into my first scratches of his chest. He shivers as I lower my face to him, but he lifts his cold mouselike nose and bumps mine with it—hello in dog terms—the briefest exchange of information.

He had a stench when he was first rescued, but the shave-down and good care have helped. He now smells sweetly of dog. The ear infection seems to be gone.

He's been a little more awake today, Tricia says, but he's still not eating well. They have tried all kinds of wet dog food, mixing and mashing and offering it to him in a bowl and by hand. He not only refused it but definitively turned away. He's taken some water but is just not interested in food. Earlier this morning, worried that he hadn't eaten anything much in three days, Tricia finally got him to take a little cat food by hand. Ollie didn't seem to be able to find it in a bowl. She's concerned. Could it be possible he's somehow lost his sense of smell too?

It is all a concern. Ollie is already far too thin. We make a run to Petco for something—anything—that might tempt him. We try him on new dog food. Not interested. We try him again on the cat food. Better, but not much. He takes a trembling lap or two and turns away. He puts his nose to a water dish and snitzes. Ollie, Ollie, Ollie. We can call the shelter to see what they were feeding him, because obviously he was eating something there, but getting a response by phone is no easy thing.

Despite the eating problem, the dog we are seeing today is really getting around. His gait is curious—a combination of blind caution and arthritic joints—and he wanders on light feet, lifting them high in a kind of circus-horse prance. Now he moves carefully across the kitchen and into the TV room, bumping lightly into furniture and redirecting himself so easily that I get the sense he has managed his blindness a long time. Several of the cats watch him from the sidelines, but Cesar trails him. Once, when Ollie turns while Cesar stays in place, they connect. It's a little like a chest bump between athletes, a little like a Three Stooges head bang, but after the first startled twitch, neither seems worried about it. Ollie continues his exploration, Cesar shadowing his progress.

Ollie is a brighter dog that evening, too, when friends come over for a spontaneous dinner party. Shauna and three other actors are there, and at one point or another, Ollie ends up in everyone's arms, even taking a little cat food from two of the guests. He is very much the center of attention, and the little dog takes to it well, despite the trembling. He is happy to be held by everyone in turn, his head lifted and his sweet face alert.

It's when the Chinese food is delivered that we get a clue to the mystery of Ollie. Lying in a cat bed near the kitchen, he's napping hard when the food comes in. It takes only a few seconds before his head pops up and his nose begins to work at the scent of stir-fried everything. He springs up from the bed and heads straight for the group of us assembled around the kitchen island, his tail wagging wildly.

The good news: There's nothing wrong with Ollie's sense of smell. The bad news: Ollie's interest says clearly that he's been fed people food. I'm guessing maybe exclusively people food. That would explain the diminished weight from his time in the shelter and the preference for cat food now, with its higher protein and fat content. As interesting as this revelation is, eating human food is not a habit Ollie can continue, and none of us likes to think what Chinese food might do to a stressed, exhausted dog's constitution. So Ollie doesn't get any Chinese food, though he certainly asks for it. He weaves his way to me and stretches up his paws on my knee in petition, patting my leg with one forepaw and then the other, the practiced movements of his past life with someone else — who must have loved him and may certainly have spoiled him with kung pao chicken. Whoever she was, whatever happened, I can't help but honor the woman who must have once responded to that approach and light double-touch. I put my index finger to his little paw, and we hold there a minute before he drops to the floor.

Later, Ollie is cradled through the dessert course and the after-dinner drinks. He dozes companionably in arms and seems to most relish being held on the hearth. Heat-seeking, he stretches out his body and tilts his spine toward the fire, flexing his toes. For a time, his trembling stops.

Later that night I carry him down to the guest suite and offer him more food. I call the mash-up of dog and cat food heavenly hash — selling it to his unhearing ears, offering it by hand. He turns away and away, and away again. *Rrrrr,* he mutters. He takes some water and turns away from food again. When I put him in the cat bed Tricia and Jonathan have given him, next to a gifted red teddy bear he seems to respond to, Ollie sinks down into the warmth of it, rolling over and rubbing his nose with his forepaws. He's comfortable and happy. He will sleep through the night without a sound. We will leave for Dallas tomorrow.

We are up early — I because I'm still on Texas time; Ollie because among his virtues is excellent housebreaking, and he has risen from his fuzzy bed and given me a set of grumbles that tells me he's ready to go out. Ollie's timing is excellent. Jonathan is leaving for work and was hoping for a chance to say goodbye to Ollie but hadn't wanted to wake me. Tricia joins us too. I hand Ollie into Jonathan's arms, and he and Ollie and Tricia have a moment together. It's a brief, affectionate interaction on all parts. I'm convinced that Ollie has begun to identify the scents of the humans who rescued him. Just before the transfer from my arms to theirs, I saw the tiny triple head-pop of a dog tagging scent. *I know this one ... and this one.* Ollie leans into Jonathan's hand for the scratching.

I turn away reflexively from their private exchange. I've been that intermediate person in the rescue of a dog bound for somewhere else, and I've loved those dogs intensely even in that short time. I can't speak to Jonathan's feelings, but I'm glad he wanted to say goodbye and that Ollie had the chance for one last interaction. That late-night vet trip they took together had made all the difference to a diminishing dog. I turn away, humbled, as I'm always humbled by how far some people will go for the good of a fellow creature in trouble. Just because they can.

First class. Ollie T is flying back first class, *because that's the way he rolls,* I post on Facebook, but actually because I think he'll travel more comfortably there. I flew out in coach but am dropping some frequent-flier miles to give him an upgrade. I want more under-seat room for his carrier and the opportunity for him to get onto and off the plane first. For a dog that came from very little, Ollie will fly to Dallas well accessorized. He is traveling in the sock sweater, accompanied by his red teddy bear, and is wearing the collar and the Thundershirt anxiety wrap Tricia bought him, in case plane travel makes him nervous. Tricia and I have somehow crammed the bed she and Jonathan gave him into my luggage, so when he gets to Dallas, he'll have their scent and the comfort of objects he knows.

Tricia and I part at the airport — a suitcase, a hug over a carrier full of Ollie, and a flash of a wave goodbye. I have gracelessly babbled my thanks to her on the way to LAX in the car, hoping to get out that it would be a fine world if every unwanted animal were thrown a lifeline like Ollie's. She, Jonathan, Shauna, and Mark had been so good to him. A week ago, none of us knew he even existed.

LAX takes us in, feeds us through ticketing and TSA, where agents do a double take over the creature in my arms. I have shucked shoes, jacket, cell phone, and keys. Ollie has come out of his carrier with his ears straight up, which gives his mouse face a somewhat shocked, indignant expression. *Yikes!* he barks as he comes out of the carrier. "Dude!" says a teenager behind me. Those ears, that single bark, carry the day. The agents laugh and speed us gently through the process, one even helping me negotiate the old dog back into the carrier afterward. The agent looks wonderingly at Ollie's ears, which obligingly swing back out of the way as we zip the carrier closed.

Today's not the best day for pet travel. Certainly not the best day for senior-rescue-pet travel. The gate area is hot and congested. Something has burned at one fast-food stand or another, and the scent of grease and scorched bread makes it seem hotter still. Every seat here is taken. It's spring break, and groups of excited teenagers rush back and forth on missions of their own that they must do in hyperdrive, squealing and pushing one another out of the chairs they've claimed.

The frequent fliers sigh. Many of the adults isolate themselves from the hubbub, earbuds in place, sinking deeply into books or laptops. Against the wall, a few frantic people are obviously on their cell phones. Each has his head bowed and one hand cupped to the Bluetooth ear, and they all alternately pace and cock their heads sideways to glare at the overloud teenagers.

Nothing about any of this seems to disturb Ollie. Perhaps he is fortunate to be blind and deaf in this moment, but I know there's plenty of uproar — movement, scent — for him too, jostling along in the carrier I try to handle gently but that probably rides rough. There is nowhere to sit. A number of people cruise the seating area like Christmas shoppers on the prowl for a parking space, and for a time I lean against a wall with Ollie's carrier tucked behind my legs to prevent him from getting trampled. When a young man a few sections away gets a gate change, he calls me over and offers me his seat. He likes dogs, he says. He's been watching me protect this little one, and he says both of us look like we need some protecting today.

A gate agent makes an announcement that only a few understand, but news ripples across the area. Our flight is delayed. Adults groan; the teenagers swear fashionably; and in his carrier, Ollie rests his chin on a forepaw. Nothing to do now but wait, and the unknowing little dog seems more come-what-may than any of us.

The howl of an angry child approaches from a distance. I can hear him long before I can see him — a series of escalating, wordless shrieks that are powerful enough I can track the course of his progress beside his mother by a shifty parting of the crowd, the ripple of heads shaking as the two make their way toward a cluster of gates.

"Oh no," says one young businessman.

An older lady murmurs, "That way . . . go that way . . . ," willing the pair to the gate opposite.

The whole body of waiting passengers is watching, and I see faces grow apprehensive and then relieved as the pair pass one gate and head for another. Embarrassed laughter follows. No one wants this sticky, screaming, flailing starfish of a kid on his plane, and when mother and son arrive at our area looking for a seat and finding no friendly faces, we recognize the inevitable. This deeply unhappy child and exhausted mother are booked on this flight to Dallas. The young woman appears to be in her late twenties. Her hair is slipping out of its ponytail. Her face is tearstained. The little boy is a preschooler, maybe four or five. She moves toward the wall where I had been standing, and something in her weary resignation makes me pop up out of my gifted seat with a speed that rocks Ollie in his carrier.

*Rrrrr,* I hear him mutter as I call them over to take the seat.

*Rrrrr,* he mutters again as they sit, son on mother's lap, and I settle Ollie and myself on the floor, wedging his carrier between me and the luggage.

The child is still screaming about everything. He is overtired, over-stimulated, "having a meltdown," his mother says to me. Her father just died, and on such short notice, she had no one to turn to, nowhere safe to leave her son. "He has special needs. He doesn't travel well," she says, but she had no choice. The writhing boy is angrier on her lap, and he's about to trade screams for kicking when he briefly takes in a breath and in that moment hears Ollie mutter *Rrrrr* inside the carrier. The boy's eyes widen, and his mother takes advantage of the moment: "She has a dog. The lady has a dog!"

The child leans forward on his mother's lap, and I unzip the case to give him a look. Feeling the new air above him, Ollie springs up sudden as a popped jack-in-the-box, all mouse nose and upstanding ears. The child tilts back into his mother's arms, then wiggles a little. Down! He wants down!

"Can he meet him?" his mother asks.

"Yes. He can meet him *gently*," I answer. I hope she hears the vocal italics. I hope he hears it too. I've got my eyes on her son. I've seen the effect that calm, elderly dogs can have on children, but I'm watching this child; his strength and tantrums could be bad news.

The little boy approaches, crouches, reaches forward. Like many children, his movements are big and broad. Ollie can't see that outstretched hand waving toward him, the small palm likely to smack his head.

"Gently," I repeat to the boy. "He can't see. He can't hear. He is much, much older than you. How tiny can you make your touch?" I show him, stroking Ollie's chest with a fingertip.

I can see all kinds of conflict in the boy's expression: impatience, curiosity, want, and the smallest light that could be tenderness. He extends a forefinger to Ollie's face.

Ollie's ears come up higher. He scents the child and leans in for the connection, bumping his small, cold nose against his hand. His tail *thump-thump*s the side of the carrier. Many small dogs are frightened of children, but Ollie seems to recognize and like them.

Or at least, he likes this one. From the side of his almost-toothless mouth, he gives the boy a touch of his pink tongue.

"Ohhh," the boy says on an inhale, a small, wondering sound that makes the rest of us quiet too. He continues to scratch gently — Ollie's chest, cheek, shoulders, back. I show the child how much Ollie likes having his bunny ears rubbed.

We sit together for over an hour in the gate area. Mother, child, dog, and I, in the center of a shuffle of passengers in the surrounding seats. The boy has grown quiet. He is staring at the dog. The child has let go all his fury, and I can see how tired he is, the dark half-moon circles beneath his eyes. With a glance at me, a half request for permission, he scoots closer and puts his hand in Ollie's carrier. Sleepy Ollie rests his chin in the boy's open palm.

We are all surprised when boarding is called. How did that happen? Our plane has somehow come in and offloaded unremarked.

"Here we go," I say to the young woman and her boy.

"Here we go," she repeats. A bystander steadies the duffel on the mother's shoulder as she lifts up her son.

Senior dogs are often victims of a special kind of neglect. Rescuers grimly recognize two periods each year — the Dump-Your-Old-Dog Days of the winter holiday season and of summertime — when owners grow suddenly weary of a dog whose age or health or general inconvenience has simply gone on, for them, too long. The owners want freedom to travel without the cost of boarding an aged family pet. They surrender the dogs to shelters or rescues. Sometimes the dogs aren't even that lucky and are "set free" to fend for themselves.

Thus, seniors get dumped in the country. Seniors are let out front doors and never acknowledged again. When seniors are crippled, blind, or deaf — or any combination of the above — their chances are even worse. These special-needs seniors are the hardship cases volunteers often beg to get help for. They just need, the pleas say, a place to comfortably and happily live out the rest of their lives.

A number of these special dogs have been in my home or passed through my hands as a transporter in the years I've been rescuing dogs. I've never regretted choosing a single one of them. All have had health issues. Most have lost sight or hearing or both. Yet old age has given many of them a particular grace. Seniors have the ability to be still and savor the moment. They are full of life in a lower key.

I still miss Scuppy, an elderly blind and deaf Pomeranian that was let out on the streets to wander after his owner died, mercifully picked up by animal control. Scuppy was so old that the shelter that held him believed he'd make it only another few months. But he was still eating, still responsive and happy, and they made every effort to find him a home for those last days. He had just turned twenty-one when I took him in. He would live another two years.

Scuppy was a miracle boy. Good-natured, outgoing, and extremely social, Scuppy frequently participated with the search team when we did safety presentations for children. He calmed even sugared-up, fractious classrooms full of kids on the brink of vacation. His sightless serenity had an effect on them. They enjoyed all the search dogs, young and old alike, but the children would wait in long lines to pet Scuppy, who was five times older than they were and who could know them best by the scent on their palms and the gentlest hello they could make by touch. They found that idea magical.

Ollie possesses many of Scuppy's qualities. Even though he is small, fragile, blind, and deaf, there is a confidence and easy sociability about him that might make him a marvelous visiting dog in animal-assisted therapies. Everyone smiles when meeting Ollie. With his button eyes, mouse nose, and rabbit ears, he's got appeal in spades. His behavior suggests that somewhere in his history, he was a very social lap dog, exposed to adults, children, cats, and other dogs — all good.

After he rests and settles in, after we build up his strength, Ollie will get the chance to train as a therapy dog. He's had a recent hard go, and it's an iffy proposition, but if that dinner-party socialization and the gentle interim at the airport were hints to Ollie's nature, I think he might enjoy the opportunity to be adored by strangers.

There's a lot of maybe in this. We'll need an all-clear from the vet. I need to see Ollie's anxious trembling stop. Ignoring dropped food in schools and nursing homes is a must for visiting dogs, so Ollie needs to train away from people food. The obedience commands that are part of therapy etiquette will be the most challenging. How do you teach a blind and deaf dog Sit, Down, Stay, and Leave It? Ollie is so social and polite, he might well have learned traditional obedience commands when he was younger — but how do I communicate with him now?

By touch?

I'm fascinated with the prospect. Touch is what he knows of us, and it's through touch and scent that he makes sense of us all. I always greet him with three scratches on the shoulder and a kiss on the head. Ollie is quick. Not long after I bring him home, he wags immediately after that ritual. He stops trembling when picked up shortly after that.

Within a month, he has associated that with the scent of me and, nose lifted, begins to find me in the yard. *Rrrrr,* he says, like the gunning of the smallest motorcycle ever, and he raises up on his back legs, wagging, wanting to be picked up. Somehow he's figured out that I, the coffee-and-chocolate-and-human-scented person bearing three scratches and a kiss, bring him only good things: a meal, a treat, a bed, an outing. Ollie's days adrift are over.

In this new confidence, we begin to craft the dialogue. A scratch on the ribs for Sit. A scratch on the chest for Down. A long stroke of the back for Stay. A reward of peanut butter for all.

Ollie is already touch acute. The sighted dogs watch him thoughtfully. He instructs us all. Far from being withdrawn because of his condition, he's an old hand at navigation through darkness. In his careful progress through a new space, he never collides hard with anything — the slightest bump redirects his movement. Through scent, vibration, or an innate sense of time, in the first week after he joins us, he begins to arrive in the kitchen for meals with the others, and I notice that he rarely bumps into another dog in that space. He often stations himself beside Puzzle, who volunteers a Down in the kitchen at mealtimes, and it's possible that Ollie has learned the scent and the size of her. There are other tangibles. He marks thresholds by a change of air current. A new scent is enough to make him pause in a doorway. It takes a while, but in his measured way, he learns the house and engages with every one of us.

I'm moved by that cheerful spirit stretching out for adventure on his own terms, despite his frailties. Beautiful Ollie has something worth sharing. Gregarious Ollie seems up for the party of it too.

# 25

THE WOMAN I SEE often at a local café says she's ready to get another dog, she's pretty sure. She thinks she'd like a dog she can train for therapy. She lost her tiny ancient poodle two years ago, a good dog friend that saw her through divorce and job loss, and for the longest time, she couldn't imagine having another. The thought of getting a new dog still feels disloyal and, more than that, frightening. She is anxious about ever growing that attached again. But her elderly father is struggling with dementia, she says, and her tiny poodle had often been a lifeline out of that haze for the two of them. Since the poodle has been gone, her father has seemed to retreat into a farther place. He no longer speaks, and it is for her father — and maybe a little, she admits, for herself — that she's decided it's time to look for another dog.

She wants a small dog that's easy to carry. She wants a mature dog that is housebroken. She'd like a dog that doesn't come with a lot of baggage from trauma, because she wants to take the dog into her father's care facility, which allows well-mannered visiting pets. Because we've gotten to know each other over coffee and egg sandwiches, she knows that I work in rescue with a focus on dogs in therapeutic uses. "Therapy Light," she calls the job she'd like to do with her intended new friend, because they'll visit only the nursing home where her father resides. But she'd like to share this new dog with anyone there who would like the visit. She has no idea what to look for in her new pet, and she wants some help.

"Keep an eye out?" she asks me. "Mature, polite, friendly. Small." She gestures the shape and size of a bread loaf. "About this big." She wonders if she's being too specific. She wonders if there's much hope for finding such a dog.

"Would a senior be okay?"

"Senior's okay."

"Are these your only requirements?

"That's it. I'm easy." She shapes the bread loaf again. She pops on her cell phone to show me a picture of the poodle she still mourns, a bran muffin–colored little creature gazing thoughtfully through a window. "This gives you an idea of size," she says, and she looks down at the picture a moment before clicking the cell phone off again.

Senior, polite, friendly, small. There are many, many dogs in shelters and rescue that fit the profile, and I tell her so. Does she want to go with me to meet some of them? She does not. She says she loves the idea of rescue, but shelters confuse and depress her. Can I find some possibilities and let her know? She loves my story of tuneful Jasper, who has since found a home as comforter to a seriously ill child. She shakes her head sympathetically over the hardships of Jasper's double rescue, laughs at the image of him whistling. Perfect! she says. Only, for her, Jasper might have been too big.

I agree to do the preliminary search, but I warn her that small dogs in shelters can move quickly — especially the kind of dog she's seeking. There may not be much time between spotting a strong candidate for her needs and that dog getting adopted by someone else. More horribly, sometimes crowded shelters just don't have much time and space to give, and if she finds a dog she thinks she wants, he might be put down before she can decide. Be prepared to act quickly! I tell her. We exchange e-mail addresses and cell-phone numbers. I head out the door hopeful that maybe we'll make a great connection here — a woman ready to heal, a dog ready for love, and a father who might respond to the touch of both.

I've also got misgivings I can't really identify. Nothing to do with the appropriateness of the woman as a pet owner, her home as a haven, or the safety of the coming dog. But what is it? There was something early in our discussion, when I brought out a picture of Mizzen and mentioned her good health and her cheerful personality. I described how good she is with the elderly. Yes, she has flights of silly, but they're funny rather than annoying. Would this be the kind of dog she might be seeking? Would she at least like to see if her father responds well to Mizzen? The woman's face changed from interest to wariness. She shook her head.

Oh no, she said, she couldn't take my dog.

Mizzen is available for adoption, I tell her. She prefers humans to any other living creature; she is now therapy trained and is a perfect candidate for nursing-home visits.

Oh no. She shook her head again. Not this one.

Mizzen didn't appeal to her. Fair enough. It's important to me that Mizzen is keenly wanted. It's important to me to find this woman a dog she'll keenly want.

I tell her I will look for other dogs, and I do. Mature, friendly, small.

Over the next weeks, I find a housebroken, eight-year-old smoky-gray-and-white Lhasa apso mix that will fetch and roll over. This guy is all engagement. He is dog-friendly, even cat-friendly. He's a lovely dog, a strong candidate in a small shelter with limited resources. Would my friend like to meet him?

No, she says after looking at a picture. A dog like that should go to a family that *has* cats. It's so hard to find cat-friendly dogs.

It seems an odd objection to me.

"I'm not sure many people visit this shelter," I tell her.

"No," she says. "Not this one. I have light furniture." She leaves it at that.

She later declines meeting a calm, black-and-tan Chihuahua on the basis of breed ("I've never liked Chihuahuas"), a Yorkie mix on the basis of size ("There is such a thing as too small"), a dachshund cross with visible cataracts ("I have a step down onto the patio"), and a small white dog with a jellybean-shaped spot on his forehead ("This isn't a good week"). She says she appreciates all I've been doing. She knows I'm giving this a lot of time. It's not that she doesn't want a dog. She is very sorry none of these have been right. But they just haven't.

Not long afterward, she perks briefly over a poodle I've found in a facility very local to her, until I tell her the poodle is male and particol-ored black and white.

Not light brown? she asks.

Not light brown. But mature and friendly. Small. Would certainly fit in a breadbox. This is delicate business, and while I want to give her an honest assessment of the little poodle, I'm trying not to push. Something tells me that no dog is ever going to be right. She is looking for the dog she lost, or maybe she's not looking for any dog at all. Her hands are shaking. She hesitates, hedges, says she will call me about it later. She doesn't. She dashes out in a hurry when I walk into the café a few days after. And then she seems to avoid the topic of dogs altogether. Our communication shifts to distant smiles, then waves. The common ground is lost, and we become relative strangers again.

I think about her often. Perhaps the search for a new dog reflects another stage of her grieving, and that must be doubly hard in the fading light of her father. Perhaps *no* to the new dog was the best answer. I think of what it must be like for her, spending weeknights alone in the home she'd shared with her good dog friend and weekends on the road between here and the place where her father drifts, where she wishes her dog back for both of them. She said she sometimes sits and tries to see the world through his eyes, waiting with him for something that no longer has a name.

The Great Cheese Leave It was tough going for Jake Piper, but public-access training at one local restaurant is greater torture still. Jake knows the space well. He's practically grown up with its cool floors, friendly patrons, and delicious scent of fajitas and bacon-wrapped cheeseburgers. While all that has tempted him before, he's trained past most of its temptations. He still finds one sweet-voiced young waitress the hardest of all to resist. She's a favorite among the café regulars, a pretty, dark-eyed mainstay, taking orders and repeating them from memory, setting accurate orders down with a flourish even for parties of eight or more. Though she might be in her twenties, or possibly older, her soft voice has never matured. It's high-pitched — the breathy, girlish sound of a preadolescent — and that, coupled with her wide-eyed, affectionate glance at Jake when he enters, is enough for him to cast all his discipline to the wind.

This is precisely why we go there. Sociable Jake has almost mastered self-control. He can be trusted to pass men, women, children, toddlers, and babies without a pull on the lead that needs correcting. But this waitress is his Achilles' heel. I've seen him break his Sit to greet her and refuse a Down in order to lean nearby, gazing upward with mooncalf eyes. He knows better. I've seen him be sly. While Jake will hold a nice tuck under the table for all the other wait staff, I've seen him stretch and shiver himself out from under it, inch by inch, so that over the course of a dinner, he manages to get near enough to maybe touch her. O precious proximity! O lilting voice! "Jake Piper, tuck under," I correct him, and he retreats beneath the table attempting to look surprised. If Jake Piper can manage to stay discreet through five restaurant meals served by the hands of this goddess, if he can stay under the table, it'll go a long way toward proving his self-discipline.

At first the young woman didn't really understand there was a problem. The nice white dog was here again, the nice white dog learning to be in service, and it took other staff members and patrons to point out friendly Jake's particular love-struck expression, the fact that his head raises when she approaches, eyes turning away from me, his partner, to follow her.

"How can I help?" she asks me. Just four words in that soft voice are enough that I can feel Jake quivering under the table.

"Avoid addressing him at all and, if he turns to you, avoid looking him in the eyes. He's got to give you up to concentrate on his partner." I don't mention that it's the sound of her he's most overcome by. I can't ask the woman to contract laryngitis or take up a two-pack-a-day smoking habit so that my dog will love her less. It's our job to master this, and on Jake's prep list for the Public Access Test, I make a note: I want his restaurant testing at a table and a time when he can prove he will resist her.

I am so busy training Jake Piper that I fail to notice another woman who's eating him up with her eyes. A retiree that we've passed often when we trained downtown on Mondays, she has a window-shopping ritual and then a break for iced tea afterward. She is a slight, composed figure, neatly dressed; there is a wistfulness in her expression that I've seen but never understood until the day she introduces herself from the next table and tells me she has a husband with advanced Alzheimer's whom she cares for at home every day except Monday. She says gently that he isn't responsible. He can't be left. Their children have provided some help in the house every day, but Mondays are "her" days, she tells me. The children wanted to make sure she got out of the house. So on Mondays she dutifully gets out of the house.

She looks across the table at me and then down at her ringed hand. The thing about getting out on Mondays is figuring out what to do with the time. What do you do when your children are grown, your income is fixed, and the man at the center of your life knows you a little less each day? She shakes her head and says shopping has never been an answer for her, and it's not an answer now. She's busy getting rid of stuff. She certainly doesn't need more. She looks in shop windows and tries to remember the last time she wanted something enough to buy it.

At least here, downtown, she can turn to people-watching. Sometimes it's good to be in the center of the activity of others. If you come to a place enough, you can feel a little part of all the things going on, good and bad. Shops opening, shops closing, people getting hired or losing jobs. Kids getting married and having babies. Or having babies and getting married. Or not. She winks at me and laughs a little. It's not proper to pay too much attention, but this is how she proves to herself that not everything is *over*. When people recognize her and call her by name, she feels relevant. It's good to be a regular and to be missed if you're not there.

She's very attracted to Jake. She heard me tell that young woman not to distract him by giving him attention, and so she doesn't either, but it's hard. He is so good, she says. So good! When her children were younger, the family had a white dog. He partly raised the kids when her husband was in Vietnam. That dog always kept an eye out. He chased off an intruder once. But he was a kind dog. Kind. He lived a long time. She speaks her white dog's name with a little pause before and after it, like an amen. You can love a dog too much, she says. He broke their hearts. They never had another.

But she quite likes Jake Piper, and she wonders if sometime she might be able to pet him. She's seen the vest, and she can read Don't Pet. She hates to ask, but she doesn't know the protocol. Do these dogs ever go off duty? Can she and Jake Piper ever be introduced?

They can. Certainly they can, but not here in the restaurant, where any service dog should be vested and on duty. I ask about her Mondays. I wonder if there might be a time she'd like to participate in Jake's public-access training. I don't want to impose, but there are test items where she could be very helpful. Jake must be approached for out-of-vest petting and receive attention with polite grace. There's a test item requiring that a stranger take him on-lead and walk him away from his partner. I'd love to see him walk with her. I'd love to see him perform well on other test items under her direction. Would she like to do that?

She would. She'd like to walk him, and she'd like to pet him. She'd like to do anything that might help Jake learn. She's got time now. She's got most Mondays, she says. Mondays are her days, remember, but she's done enough shopping. She'd be honored to give part of her Mondays to Jake. We exchange phone numbers, and when we step outside and Jake's vest comes off, she sits on a bench and wraps her arms around the barrel chest of him, bowing her head to his back.

As we prepare for Jake's Public Access Test over the following month, Jake sometimes on-lead with me, sometimes on-lead with her, our friend warms to the idea of having another dog. She wonders if it was selfish of the family to shut themselves down after loss. They had a good home and love to give, and all these years — there were so many needy dogs out there. In time, she asks if a dog could be trained to find her husband when he wandered in the house or yard. Could the dog also be taught to bark when her husband tries to unlatch their locked doors?

Yes, I say, on both counts. There are service dogs out doing just that job now — with seniors and with autistic children. If she's interested, I could help her train one. She hated to ask, she tells me, she hated to impose. But that would be wonderful. She imagines what it would be like to have a dog friend in the house again — a dog friend that could help her husband and her family in important ways.

Threading Jake Piper through an antiques store, she lets the matter drop. But the next week she asks if I might help her find a dog. A good dog just like Jake, she says, but so we're not disloyal to our last one, any color but white.

If you want to test temptation for a would-be service dog, few places rival PetSmart. On a busy Saturday afternoon, that place has it all: kids, cats, birds, fish, noise, toys in reaching distance, and everywhere, drifting over everything, the sweet scent of kibble. It's not the first place I'd go to train a dog, but for a dog approaching the Canine Good Citizen or Public Access Test, it can build confidence or clearly show where more training is needed.

I liken training a dog for the Public Access Test at PetSmart to having a kid take a math exam at the circus. Woo-hoo! Yet here we are, Jake Piper and I, because he's pretty much proven he can hold the Down/Stay beneath a restaurant table and pass a toddler's dropped ice cream without struggle. He's visited busy Dallas downtown and seems unworried by the general screech and roar of the city. He's ignored passing dogs. He's sat nicely for petting (vest off) and patiently allowed a groomer to examine an ear. He's held a Down while a shopping cart rattled by just inches from his tail. He took all of this in with no more than a bemused expression, the ears alone moving randomly during each event. None of it seemed to upset him.

Today's work is being done on an unusually busy day. A couple of animal-rescue groups are on premises — meeting and greeting with their hopeful adoption candidates in bright bandannas. It's a perfect day for Jake to show me he can calmly pass other dogs. It's a perfect day for the rescue group to show that their adoptables are friendly and social. One splendid senior black Lab mix, so gray about the muzzle that his head appears white, calmly works the aisles with a volunteer. He's wearing a Donation vest with a little arrow embroidered on the pocket rather than an Adopt Me vest with a little heart. He and Jake pass each other with no more than a slight head bob. They sit while the volunteer and I talk, the old Lab settling in with a little arthritic groan, Jake rolled back slightly on his haunches, tail swishing idly across the floor.

The volunteer says it's been a pretty good day today, though everyone wants the light-colored puppies first, and the black dogs are getting little interest, which often seems to be the way of it. Black-dog syndrome, the condition of being overlooked due to color, seems to be in play here. There are many guesses as to why black-dog syndrome exists — black dogs are more common; they don't photograph well; there are superstitions surrounding them (similar to those surrounding black cats); some feel they are just too plain, and others that they look too mean. The rescuer says that for a time after J. K. Rowling — God bless her — created Sirius Black, a character who could shape-shift into a dog, the black dogs in rescue got a break, even a little advantage, maybe. Black dogs in that period *moved*. There are probably hundreds of Sirius Blacks out there now, some of them one-hundred-pound rottie mixes, others inkblot-sized Chihuahuas that top out at three pounds soaking wet.

Max here — she gestures to the elderly Lab — has been a foster for five years. He is the best dog ever, came in already knowing Heel, Sit, and Down, friendly to everyone and everything, and no one ever seriously considered him. Now Max is twelve years old, and the rescue has accepted that he is unadoptable and stopped trying. He is their mascot, their Donation dog, and though he's had a loving home in foster for a long, long time, it's a shame, because he would have made some family with kids or someone who wanted a quiet, mellow friend very, very happy.

I can't believe no one ever considered this boy.

"Well," the volunteer says, "one guy did that I remember, but then he saw that Max has a kink in his tail and didn't want him. Seriously, it was that lame an excuse." She reaches down and affectionately feels for the marble-sized knot that made Max a no-deal. She says it may have been a stroke of luck for the dog. A guy who got worked up over a tail kink might have been a real jerk.

I tell her a little about my work with Jake Piper and the firsthand research into training a psychiatric service dog. She likes Jake's soft expression. She likes the way he met Max — friendly, but not rambunctious. I take off his vest, and when she pets him, she laughs at the way Jake takes affection — the flash of shy smile that some dogs do, little pearl incisors showing, leaning his head into her hand, folding his ears, tail *thump-thump*ing rapidly like a puppy's.

Psych dogs. She has never heard of them. She says depression runs in her family. Her grandmother had it. Her father was a veteran who came back from Vietnam with it. Sometimes he was better. Sometimes he was worse. He was never quite himself after he came home. A "woman's weakness," in those days; a man didn't admit to depression. She has one arm around Jake Piper and one arm around Max and says her father self-medicated by drinking a lot. She shakes her head and says he called it "killing the black dog."

• • •

"She was absolutely not the dog I imagined. When I met the transport truck with Mary, the man was taking the dogs out one at a time. He yelled, 'Who has the black Doberman cross?' I was just standing there, because at the time we didn't think that's what she was. And Mary said, 'Oh, that's Pam's dog,' and I was shocked. I said, 'Whaaat?' And out comes this dog, so scrawny. It made her look really tall because she was so skinny. She came out of the trailer, and she was just flying in the air, no feet on the ground, just bounding on the leash, and I thought, *Oh my God, what did I just do?*"

Pam is describing her first meeting with Babe, a rescue pulled from a shelter in South Carolina. Babe was just around a year old. She was mostly black, a rottweiler-hound cross that looked deceptively like a Doberman. Like so many others, Babe was one of those dogs typically bypassed in shelters. Field and power-breed mixes often are — for many would-be adopters, these dogs are too big, too high energy, and too dark in color, and, among many other similar dogs in the shelter, they are difficult to tell apart.

Pam had been looking to rescue a dog for a while, and she'd been looking to find the right dog from that slighted population. She had another dog and knew her household; she'd been thinking a black Lab or Lab mix, maybe three to five years old. She says she always wants to get it right when she adopts, so she takes her time. For eight months or more, she looked at hundreds of photographs and read the accompanying bios — and in a sea of black dogs, this one dog had something special. "Babe had the softest eyes," Pam says. "The minute I looked at those eyes, I knew this was the one."

Mary, who runs a rescue group in the Northeast, had originally gotten a plea about Babe from two volunteers in the South Carolina shelter. They too had seen something special in her. Anxiety and boundless energy notwithstanding, Babe was extraordinary with kids, so exceptional that the volunteers urged Mary to list her on her rescue site. Requests like these are often tinged with urgency. Mary recognized that here. Though she wasn't sure this dog's description was going to attract many adopters — all the wrong pings of color, size, and deportment — she listed Babe through her rescue anyway, and Mary's rescue partner Annette went to the Aiken Animal Shelter to photograph her.

Great photographs can make a difference for so many of these dogs.

It's not always an easy capture in the shelter environment, and dark dogs are harder still, but Annette is particularly good at it, Mary says. Annette caught the dog's intelligence, beauty, and, as Pam put it, "her soft, kind eyes." Annette's photograph made the difference, and in a much shorter time than Mary had predicted, along came adoption interest — from Pam.

Still, Mary says she wondered during that phone call if Pam had the right dog. There's always risk in rescue. Mary wonders every time if a dog is going to end up back at her house in a week, so she tries for full disclosure. She wants people to know as much as possible about what they're getting into, who they're bringing home. Though this dog had no aggression issues, she was super-anxious and energetic. Mary thought she knew what Pam was looking for, and this dog was wide of that mark. "Are you sure?" she asked Pam before the adoption took place. "Are you sure?"

Pam was sure then. She seemed to stay sure, Mary says, even at that first meeting. Any uncertainties she may have felt when Babe levitated out of the trailer, Pam managed to keep to herself. Mary was impressed. Pam seemed as calm as Babe was hyper. Pam saw something in the long-legged, kiting creature; she somehow knew the dog was going to turn out fine.

Struggling to merge her preconception of the dog with the great, leaping reality of her at the time, Pam says: "I believe in commitment. I said I would take her, so I was going to take her, and within five minutes I was in love with her. Just the sweetest dog [and when she got up close], there it was — the most beautiful look in her eyes."

Pam named her Caro, a soft name to mark her new beginning.

Pam already has a therapy dog, and she wasn't really looking to adopt another to do the same work when she brought Caro home. Therapy-dog policies usually specify one dog per handler on any given visit; with two therapy dogs in the house, Pam would always need to make a decision about which dog to take, and she didn't really want that. Caro was brought into the house to be a pet.

But then Pam saw the dog's reaction to children. Pam had never seen anything quite like it. Around children, the big dog goes boneless, she says. It's like she's received some kind of drug. Besotted, devoted, name the loving adjective and it seems to apply to Caro around kids. Big kids. Little kids. She likes to listen to them; she likes to be near them. If they want her to sit, she sits. If they want her to lie in their laps for a full-body hug, she wants that too.

Pam knew she had seen something in her dog in that first photograph; she remembers a low-key reference in the write-up about "good with children," but that can mean anything, even something as minimal as a wag to a passing kid volunteer. As an experienced rescuer, she also knows that rescue dogs can hold surprises. It takes a while to figure out how much good and how much bad they carry from their past, what their triggers are. Pam socialized Caro and carefully watched her in their first months together, and while nothing seemed to faze the dog, Caro's eagerness to please and her joy beside children was a surprise to Pam. Caro was born to work therapy. It seemed a shame to let those gifts go to waste.

At first glance, others might not be so certain. There it is again, black-dog syndrome, Pam says, and it's all such a complicated read for strangers — the big dog of a power breed, the dark face whose expression is harder to decipher, the bobbed tail that wiggles where people would look for an all-out wag. Pam remembers a woman from their early days who glared across the waiting area at Pam and Caro before the training class began. More than an angry glance, the woman gave a series of hostile stares in Caro's direction. They were impossible to miss. Pam wondered what made up that history, what was provoking all that, and finally she'd had enough. She made eye contact, neutral but unapologetic, as though she welcomed anything the woman had to say.

"Do you ever get anything bad said about you because you have such a mean breed?" the woman asked.

Like so many questions, this was really a statement.

"No," said Pam. "I don't." And she let it go.

But during the course the class, which also included the woman's husband, Pam made a point of keeping Caro in the woman's sightline. Sweet, eager to learn, and in control, Caro had it in her to override prejudice, and Pam wanted to give her a chance to do it. After the class, Pam purposely walked Caro close to the woman and stopped near her as she was gathering up her things. Pam knew what was going to happen. Caro wiggled her way up to the woman and, at some sign of invitation, sat down, leaning on her, just doing her sweetness. Pam could see the change. The woman first bent over to pet her and then couldn't resist hugging Caro, saying, "Oh, she's just the sweetest dog." It was a little moment, just a little thing, but it felt good to change one mind.

Pam is a thorough partner. Caro was a quick study. Caro earned her therapy-dog certification within six months of adoption. With Pam, Caro works in the Paws to Read program at a local library, a "listener" for children to read aloud to, an encouragement for anxious readers who struggle. Caro's compassion for kids is leading other places. Soon she will assist at a local child-advocacy center, a support resource while children who have been sexually abused wait before forensic interviews. Such centers strive to be child-friendly, and dogs like Caro provide distraction, encouragement, and calm.

Going from last-chance dog to goodwill ambassador is a long journey to make in a short time, especially for a dog that once literally climbed the walls of an animal shelter and exploded from the side of a horse trailer.

Mary comments on the phone:

"I didn't go into rescue thinking that this would happen. I went into this to save the lives of dogs and to match those dogs up with good families and people. But to see one of our dogs — [like] Caro in her therapy vest — it is that sense of accomplishment that wow, we had something to do with this. It's good work. It's worthy work. Pam deserved the chance to have this dog, and this dog deserved the chance to have her. It was just a great meeting of spirit and generosity and energy. She was not a dog everybody would be able to take. They would just overlook her.

"We've had it happen that way before. Someone sees a dog and says, 'That is my dog,' and he's a plain brown dog. But that doesn't happen often. And it's different from when it's a little adorable Maltese cross, and he's very cute, and someone, several someones, will quickly say, 'That is my dog,' about a little Maltese cross. And little Maltese crosses deserve good homes too. I don't try to figure it out. But when it happens for one of these dogs — for one of the black dogs or the brown dogs — that's when I think . . . ah . . . it was a very good day."

# 26

IT'S A BLUSTERY, UNCERTAIN morning when I leave Puzzle with the vet who has long cared for her. Puzzle picked up a skin infection after search work in an unfamiliar area, and she is now very sick. She is always slow to show pain, but yesterday, on the Take Me Back portion of a search exercise, she lay down on a sidewalk, put her head on her paws, and whimpered. It was only the third time I'd ever heard her cry.

What began as an annoyance, a rash across her belly, has progressed to something that I cannot seem to get control of — not with Eastern, Western, traditional, and holistic therapies; medication; food changes; flea preventive for dog, house, and yard; healing shampoos; and all kinds of prayers and psychic incantations for her by friends near and far. She rallies sometimes and always wants to go out-out-out to work, but I don't know why that's so. She hasn't been able to shake whatever this is. Now Puzzle's exhausted with itching, and it seems her inflamed skin cannot heal. Golden retrievers and itchy skin — a common misery, apparently, but it has come to a desperate pass. This morning I could not find an inch of my dog that wasn't inflamed.

The vets and I have tried every medical approach we know over the past four months. In that time, I've watched my merry girl diminish. She's lost fur, she's lost weight, she's lost joy. I've pulled her from wilderness work until we can get her well, and Puzzle still lights up briefly during urban-search training, but then she comes home and fades. Even after she has a soothing bath in oatmeal shampoo to get rid of whatever might be out there in wherever, it's scratch-scratch-scratch, often deep into the night. By day she withdraws into a fitful sleep.

The only relief we found was during a brief visit to Colorado during a late-spring snowfall, where the itching stopped and her eyes grew bright again. Now I half wonder if my dog has grown allergic to Texas. I half wonder if we'll have to move to heal her. Search-and-rescue friends in other states make offers. They can keep her in the bad seasons. They have a room, a guesthouse, an apartment where we could both come and live and work in the summer worst of it. They offer every kindness, and I consider their offers carefully. Puzzle is a young dog still, but she moves now like she is very old.

I stay with Puzzle at night, following her restless course from room to room as she tries to elude the itching. We move from my bed to the couch to the daybed to the living room floor. Puzzle paces, and I am just behind her, dragging a pillow and a quilt to lie near wherever she can manage to settle. The dogwatch, I call it with any of my sick dogs, feeling better about being nearby in case one takes a sudden turn for the terrible. We've had enough sudden turns for the terrible that go more terrible still, and I can't bear the thought of that happening to Puzzle. How bad does infected skin have to get before her other organs are affected? Like so many ill dogs, Puzzle has grown remote in her distress. Even so, she seems vaguely grateful to have me close these nights. Often when she feels me settle beside her, she extends her head to give my hand a lick.

Jake Piper, in turn, shadows both of us. Puzzle has ceased playing with him during these sick days, and he seems aware that she's unwell. He doesn't try to provoke her into play, but he lies nearby with an intent expression, watching me, watching her. He is the only dog in the house that chooses to do so. The little dogs are mildly disturbed by our restlessness — a few grumble, the new, sightless rescues raise their heads and gaze toward us like oracles, but Jake Piper alone is involved.

When the vet recommended that I bring her in so he could isolate and treat the skin infection, I hoped, as I'd hoped all the other times, that at last we'd get an answer. I felt heartsick leaving her behind, and she was bewildered by it: my beautiful, vibrant, happy working dog who, other than when she was bitten by a snake, has never been sick in her life. I cradled her in my arms and kissed her too-hot ears, and when the vet returned to take her, I went so swiftly out of the room and down the corridor and out the door that I heard one customer say to another: "She must have lost that little golden."

It helps to be busy. I can't stop myself from falling into the old routine of getting on with things, so for the next few days I make a too-long list of errands and overbook myself to distract my thoughts from a much-loved dog in shadow. The little dogs get bathed and brushed out. The house is impossibly, even obsessively cleaned.

But there's a difference. I've done enough pacing over sick dogs in the past to feel it. Jake Piper is now a figure in this waiting. He's a dog trained to attend a human partner, and he does so. He is pensive over Puzzle's absence, but he keeps me in his sightline, stays beside me nonetheless. A calendar note shows that he's due for a trip to his own vet, who's closer to home. It's just a checkup that could be done at any time in the next month. We'll do it now, I say to him. Despite his confusion, he carefully watches me lock the door. When I don't double-check it, he seems relieved for both of us. He hops in the Jeep without a backward glance. Jake seems eager to be out of the house. Inside, he never stops searching for his missing friend.

For a dog who's about to do nothing more than get a standard vet check and a toenail trim, Jake Piper is awfully pleased with himself. He seems to think any vet visit is one big party for the good white dog. He pads into the waiting room at the vet's office, gives a friendly chuff to the office cat, and wags wildly to the receptionist, but when he's asked, he drops into a polite Down/Stay at my feet and waits for whatever celebration of Jake Piper is to come. The receptionist leans over the counter and remarks on the change since she last saw him. Jake is bigger, stronger, his ears are up, and his eyes are clear, she says. And he's better behaved! We say that almost in unison. The first time he came here, he was still a little freaky over friendly young faces. I can see the wannabe-freak in his eyes even now, looking up at her smiling down, but he holds Sit, he holds it, and he doesn't budge as she coos at him, against all odds and despite the bowl of dog biscuits on the counter.

Then the glass door opens, and a man and two children enter leading a long-coated white German shepherd. She is a beauty, with dark almond-shaped eyes, upright ears, and a gracious, regal manner. She fills up the room. Everyone stares at her for an extended moment — she is that kind of dog — and then all of Jake's obedience evaporates in the flash of their connection. He gives a happy shriek. The German shepherd lets out a single bark at the same time that Jake Piper belly-crawls toward her whimpering, his ears folded and his tail whip-whipping across the linoleum floor.

"Can they meet?" asks the gentleman, ready to reel in his dog.

"They can meet," I answer, lapsing as a handler, unhappy with Jake's lapse of obedience but wondering at this. He has never responded to an unfamiliar dog this way before. Jake frog-swims over to her on his belly, remains low while she nuzzles his ears and licks his face with meticulous swipes. He squeaks and waves his paws, rolling belly-up before her. It's a playful lovefest, a lovefest interrupted when a client wrangling cats in two carriers pushes the door open with her backside. The poor woman is overladen, her purse, bra strap, and glasses slipping every which way, her cats yowling profanities from the depths of the carriers.

We separate the dogs to help her. When I call Jake back to me, he surprises me by scrambling to his feet and returning quickly, sniffing my hands, his eyes wide and his ears alternately perking and folding at the tremendous noise set up by the unhappy cats. A spiked, angry paw fishes out of one of the carriers. Jake knows cats, but this . . . Joy, wonder, curiosity, fear all play across my transparent dog's face.

The German shepherd isn't ready to let go. She is uninterested in the cats, but she strains across the floor to get back to Jake Piper, her tail waving lightly. Her owner is apologetic; she is usually not this way. The children haul away at their dog by pulling on the leash in their father's hand, and the German shepherd acquiesces, giving in to their pull even while her eyes are on Jake and his are on her. The two dogs are completely different in build, coat, ears, and tail, but their faces look so much alike in that moment of parting that I suddenly wonder if this is Jake's mother.

Such a beautiful dog. The children tell me she came from a shelter. More than a year ago, their father adds. They had in mind to get a puppy, but then they found this grown dog. My God, he says, there were so many dogs and not enough people coming in the door. They saw her big brown eyes and had to save her. Perhaps, like Jake, she too had once been abandoned. The timing of her adoption at the shelter and Jake's appearance at my house would be right.

But their dog is as elegant as mine is ungainly. The two dogs do have common features and some shared spark — but what is it? I've heard that female dogs and their puppies typically lose a sense of each other over time. It seems wildly romantic and Disney-esque to think that Jake and his mother have found each other now, to believe that on a random day by coincidence some terrible parting was healed.

What does matter is that their meeting today has given Jake Piper some happiness. It could be, of course, that Jake Piper senses something in the female dog that is similar to Puzzle, the dog that actually raised him. There are similarities between this girl and our golden, too, especially the soft eye and the quick, affectionate acceptance. Perhaps for Jake, even though this dog isn't Puzzle, she is close enough.

The German shepherd's appointment is before ours. Her name is called; the family leads her back to the examining area easily, her pretty head up, plume tail waving as she *tap-tap-tap*s across the speckled linoleum floor. Jake Piper cannot take his eyes off her. I have not asked him to. He holds his Down/Stay, but with his head resting on his paws, he watches her go.

After his checkup and nail trim, after all the praise and cookies that come to fearless, social Jake Piper back in the reception area, we head home. The Poms seem furious that he has had some adventure that they have not. This is always the way of it. Though they can no doubt smell the vet's office, which they dislike, they appear to scold Jake for getting a trip they were denied. They scrimmage around his ankles, suspiciously working the scent he's brought in. Jake ignores them, which can't be easy. He turns away from spinning Mr. Sprits'l, sidesteps Smokey, and high-steps over Mizzen, who would feed her head to him. Jake disregards all of them and immediately begins looking for Puzzle, clearing the house the way a search dog would, room by room.

"She isn't here. She isn't here," I remind him. We're told dogs live in the moment. That sounds like a virtue, until I realize Jake may have the new pain of loss each time he returns.

He clears the house and returns to me. Whether he stays close because he senses my distress or because he seeks to comfort his own, I don't know, but Jake is at my side. The third night without Puzzle brings me a series of bad dreams, a tangle of losses in the search field that will not resolve, and I wake to find Jake has crawled up into the Solace position that I taught him, his paws on my upper arm, head resting on my chest just in the place where a heart would ache. *Solace* is our command word for the task so often needed by those with panic attacks, night terrors, or sudden sinks of depression, and I taught it to Jake without ever thinking forward to a moment he might need to offer it to me.

The Poms are also anxious, and they have spun up with it. On the third day, the only way to quiet them is to give them something else to think about. Jake and the little guys get Sits and treats for goodness, and then I vest Jake up and lock the door — lock it only *once* (he is watching) — and we head out to train. I don't know what he understands, how he framed Puzzle's illness, which he could surely scent, and how he makes sense of her absence now. He is subdued where he'd normally be electric on a walk. Jake Piper steps into a Heel beside me as we turn northward from the house. The prettier walk is the other direction, but we head out along the same street Puzzle and I once took while looking for Lost signs and waiting for a word from the vet about puppy Jake.

I can see approaching weather ahead of us, a blue-gray roll of cloud with a darker sky behind it. This is the kind of weather that Puzzle and I love to work. With the wind swinging across the compass, working scent direction becomes a challenge. If she were home, and well, I would have Puzzle out now. I would have called some neighbor, asked some friend to leave a shoe out for us and to hide blocks away. Puzzle would have enjoyed trailing our volunteer missing person to the find, and then she would have brought us all home, head up, tail swishing, prideful.

But Jake is with me, bless him. His working walks have other tasks. His walks are about attentiveness and self-discipline and orientation. So we'll head north, and east, and randomly south and east and north again. We'll walk until we're tired. My intention is to wander the way any distressed person might, and carrying the weight of worry for Puzzle makes it easy.

We'll wander, and maybe at the end of that meander, Jake can lead us home. He's been closer to success with the long-distance Home command lately, though once he sidetracked us to a playground full of children, and last time he took us to a very promising backyard barbecue with a keg, a smoker, and three grills going. Apparently, to Jake, home is where the party is.

No kids on playgrounds. No steaks today. He looks up at me now and again in question. It's not Puzzle's glance to double-check my position on a search; Jake is focused on my face. Once I catch his gaze and realize I've stopped in the middle of the sidewalk and am just standing there. Scared has caught up with me; sad has too.

As one neighborhood shifts into the next, we pass more dogs than people. Some of the dogs are interested in us. They poise at fences, some wagging, some springing up as though savage, some booming happy woofs out of eager faces. Some dogs seem anxious with the sense of storm. We see a brownish Lab mix pacing on a chain in a side yard. He is a big dog, but exposed to the weather, he is reduced by every new gust of wind, has his ears down and his tail tucked. He stretches to the very end of his chain, yearning toward a front porch he cannot reach. We see dogs in crates under carports. We see inside dogs straining to see us, their paws on streaked windows. One tiny, very senior fuzzy mop appears to be on the back of a couch. She's a pretty little dog, wreathed by a quilt in what seems to be a place of honor. She has a whole picture window to herself, and though she watches us pass before her, she doesn't speak. Jake Piper is silent also. He notices all the dogs — ears perked and gait growing more sprightly in some nonverbal exchange with some of them — but he holds his walk beside me.

Somewhere along here must be the home of Mizzen's friends, the wheelchair-bound mother and her daughter. I wonder if these dogs we are passing are, like Mizzen, loved, familiar landmarks to them.

We walk absently for over an hour, sunless, sometimes heading into wind and intermittent rain. The training mind of me observes Jake Piper's focus and his loose-lead heeling; the rest of my mind is somewhere else. Maybe it's those dogs, that route lined with evident affection and neglect, and thoughts of Puzzle that make me realize where we are.

The last random turn has brought us two blocks from the house and shed where Jake Piper might have been born. Not interested in making him revisit some horror he escaped, I have never taken him back there. But he's a different dog now. Our chance meeting with the white German shepherd — Jake's eagerness followed by the calm, thoughtful goodbye — makes me wonder how much of his hard beginning Jake Piper still carries.

We turn for the house that's in the middle of that riddled neighborhood, more houses leaning, more houses razed, and just about the point I think we've somehow passed it, I realize the house and the shed that might have been Jake's are gone. The raw turning of the driveway from the street is still there, and so is the scrub line of volunteer trees along the periphery, but that's it. No house. No God-awful shed. Nothing. The demolition must have been fairly recent. A backhoe is parked where there was once a porch.

It's like a benediction. I hoot a cheer. I've never proven this place was Jake's beginning, but it's always represented his early misery anyway. The destruction of that damn shed feels like a victory.

We walk onto the land, and I take off his vest to let Jake wander. He perks. He is certainly interested. He works his nose low, the curiosity of a dog finding all kinds of smells in freshly overturned earth. Jake has always been a dog that likes to share his discoveries, and he does so now — finding things, beaming up at me with excitement: A dog's piss line near the sidewalk! a shred of Styrofoam cup! an *earthworm!* Oh, the sweet pleasure of rolling in earthworm. Nothing about Jake suggests fear or even recognition of this place. I too am looking for anything I remember here and find nothing at all.

My cell phone buzzes in my pocket. A call has come in and the phone didn't ring, or it rang, and in my thick thoughts, I somehow didn't hear it. I recognize the number: Puzzle's vet, who checks in daily. A voicemail has translated to a text. My voice-to-text service is famous for its terrible transcription, but in the mangle of this one, I think I read good news.

FUZZED UP DOING SWELL. GIDDYUP SITUATION. HOME
TOMORROW. BYEBYE.

Does this mean Puzzle's doing well and can come home tomorrow? After a crazy-making listen to the distorted voicemail, which sounds even more garbled than it spells ("Fuzzed up Phil doing swell. Giddyup bitching. Dangle home tomorrow. Bye bye") and a hold when I call back the clinic, the receptionist confirms the vet's message. Yes, Puzzle's doing much better. Her skin is still a little raw, but she's eating and bright-eyed again.

And "giddyup situation"? The receptionist has no idea, "but if it sounds like good news," she says, "I'd take it."

What does Jake Piper think of me, laughing out of nowhere, crouching down to him? He stops his exploring to sit very still with his head on my shoulder, occasionally licking my face, and he stays there while I work out the good cry that has never come easily. Jake is unsure what to make of the sound or the salt of it, so he tries all his comforting tricks: the lean, the puppy wag, the wash of my face. Caught by something odd about our posture, a passing driver slows, calls out the window of his truck: "Ma'am, are you okay?"

We're okay. Jake Piper has his eyes on me. I wave my thanks.

Recovered, we walk across the abraded property. Funny how good news changes the whole aspect of a place. Puzzle coming home makes even the grit of a demolished house and a grubby backhoe look better. Rain patters over the mud, lets up, and then falls harder. A handful of sparrows are happy in the scrub hackberry, branch-hopping, shuddering their wings in the clean of it. The earth smells sweet. Jake, who hates a bath, is crazy for rain. He snorts and capers to the end of his lead and wags his way back again, muddy to the elbows. We are soaked to the skin. We've got a long walk ahead.

Vest back on. "Home. Take me home, Jake," I say to the white dog. He leads out in confidence, as though he knows where that is.

He doesn't.

"Never mind," says Paula later. "He's still learning. Not every dog has to do every task." She encourages me to think of it this way: in Puzzle, I've got a dog to teach the finding; in Jake, I've got a dog to teach the staying found.

# 27

MONTHS LATER, AN IMPROVED dog-to-teach-the-finding nudges blinds aside to stare through the window. Puzzle's brow is furrowed. She sees me with a suitcase, and she sees Jake and Jake's vest and Jake's bag of gear all being loaded into the car. Departure is a rhythm Puzzle knows, and this is surely departure, but for years, *she's* been the dog that dashed out of the house; *she's* been the dog that leaped into a car full of gear. Today is the first time the motions of travel don't include her. It's a dark look she shoots me through the glass. Behind her, Mr. Sprits'l, who cannot get to the window, scolds. I could caption the meaning of it: *Welcome to my world! This is what it's like to be left.*

Jake and I are headed out on a road trip alone. While a service dog's Public Access Test is exacting, it doesn't take much time. A road trip like this one better evaluates a service-dog-and-handler team's readiness to navigate an unpredictable world. Long hours driving, multiple public settings, obedience and task commitment in strange places with noise, new people, unfamiliar dogs — this is where we'll evaluate Jake's constancy as a service partner across days. This is where we'll see, too, what I bring back to him. We travel for the sake of experience with change, not to a destination.

It's a proving opportunity. Some dogs that are comfortable on short trips develop intense carsickness on long ones. Some tense up with the stress of travel, and their good manners fail. On trips like this, even great dogs can get too distracted to be of service, too overstimulated to pay attention to even the simplest commands. If Jake has issues we need to address, they'll probably show up this weekend. I hope he doesn't have issues. I hope for good things — polite public presence and real commitment to the service tasks I ask him to provide — but Jake has never been on duty for so long on the move. This trip is a very big deal for a young dog beside a human feeling not altogether well.

A peculiar journey we have before us too. While we have a route and a schedule, we'll wander from it sometimes. I will try to find the unexpected beside Jake. We'll visit small airports. We'll pull over to construction sites — backhoes, dump trucks, nail guns! We'll walk past tractors in motion, if we can find them. We'll look for weird. Because more than one service dog has ultimately washed out from a fear response to black trash bags (some cower, some bark, some helplessly pee), we'll cruise random neighborhoods hoping for trash day. A long walk down blocks of black bags and the possibility of noisy trash trucks — it's not everyone's idea of a weekend holiday, but for us, it would be a solid score.

Jake needs to demonstrate his focus. Though he is an eager learner and has a willing heart, Jake is by no means a perfect dog. He has his temptations. The worst of them is probably children — sparky, fun-sized humans that he can often look straight in the face. Jake loves children, and children love him for his kindly expression and the spotted, akimbo ears, the way he pays attention and tilts his head back and forth when they giggle. Cute as it is, the mutual attraction can be a problem. Jake has learned that the Meet and Greet command means he can be petted by strangers, but he never gets that command while wearing his vest. For tempted Jake, that would send all kinds of mixed messages. It would also send mixed messages to many of the strangers we meet, who are often much confused by the petting policies of different working dogs in vests. So it's cleaner all the way around if Jake learns he must be out of the vest for petting and he must be given the Meet and Greet permission. This is even harder for Jake than Leave It. I can tell. He could throw everything over — obedience, duty, even good manners — for a towheaded toddler with a gummy bear. We won't exactly be bird-dogging for children on this trip, but kids and service dogs happen. He's had great self-control lately, but one long-term service dog trainer told me that away from home, sometimes even a great dog's discipline can go *poof!*

There have also been new indiscretions. Jake's been behaving very well in restaurants lately, tucking himself under the table, resting against my foot, and virtually disappearing during a meal. He has ignored crumbs, turned away from food flung by a nearby baby, resisted come-hither motions from servers. He has winced and buried his head against booth upholstery when a café singer with an edgy voice made my own ears hurt. But recently, as I was having a meal with a friend, a lively, animated woman, I noticed a random twitchiness about her. It got worse, that fidgeting, and when she jumped outright and looked beneath the table, I knew Jake was up to something. Good Jake, curled handsomely beneath us, was certainly being quiet and out of the way — at the same time he was licking my friend's sandaled feet.

He had never done this before — not to me, not to friends, not to strangers. Jake has never loved my feet at any time, but there was something about my friend's that he could not resist. "Leave It," he heard from me, and marked, and even shifted away when I told him to, but as the meal progressed, he returned to her in creeping measure, close enough to cold-nose her ankle in one moment, turn away as though he'd done nothing, and sneak in another kiss on the bridge of her foot minutes later.

Dog kisses get a mixed reception among my friends. Some love them, some accept them in moderation, and some don't like them at all. This friend was not a fan. Jake's kisses were unwanted, and beyond that, from a service dog, they were completely inappropriate. It will be strange on this trip to choose restaurant tables by how close someone in sandals is sitting, but I'll do it. If Jake can be tempted by feet, we'd better find out now.

My own readiness is also on trial. My strength fluctuates and sometimes I cannot feel my steps altogether, and I've come to need Jake Piper more. Recent indications suggest that in time I'll need a service dog daily. Even though I knew this might eventually happen — it's been forecast often enough — I still struggle with the idea of slowing down. Where is the clean line between denial and giving in to something you could beat? I can never find it. Claiming to like straight-up realities (and having just spent several years working with service dog handlers who have taught me how important that is), I still try to ignore plenty of my own, reading up on the body's gift for compensation, holding on to a memory of stamina I can no longer trust. Push, push, push. Some days I think I'm winning. Most days I am.

But I had a bad fall stepping out of bed one morning. The numb foot that was there might as well not have been. I crashed into a nightstand and then onto the floor so hard it left a row of bruises up my spine. I lay there awhile, looked up to Jake Piper and Puzzle looking down, and thought: *So there it is, and here we are, the service dog in training and I.*

I had thought Jake might be the dog I'd learn with, the dog to help teach others (including the dog I'd eventually need), but now I don't know. We work on every task Jake might demonstrate for future dog teams in training, psych service and otherwise, and we work equally on those he might need to do for me. We move from theory to reality, Jake's steady gait to my sometimes confident, sometimes faltering step. It's an uncomfortable, challenging time, but I've managed to evade the depression often associated with chronic illness. Doctors, expecting it, have asked. I think the work beside the dogs has made the difference, and I say so.

I tell Jake: "If this is what's coming, kiddo, I'm glad it's you beside me."

I tell myself: *Pay attention. Have adventures! Trust the dog.*

Traveling with a service dog, even a kind, obedient, socialized service dog, is far less simple than traveling alone. Even everyday choices take some forethought. His things, my things — we are double the packed gear; I expect most of our activities will take double the time too. It's a good test of us — this week my right foot and hand have gone especially numb. Stairs, uneven sidewalks, door thresholds can all be a problem. Task-trained to brace a potentially wobbly partner, Jake has recently transitioned to a stability harness. I can say "Brace" to him, and Jake will dutifully steady or counterbalance, whichever is needed. He has another command that yields a slow, steady pull when I'm going up stairs; he's as solid as a handrail. Still, there will be luggage, and I've chosen guesthouses for us where I'll have to figure out the logistics of bags, stairs, and service dog. If I have weak days, I'll have to mete out my own strength carefully. It is the thoughtful process many handlers must manage.

Though we've held off on this trip waiting for cooler weather and much cooler asphalt for Jake's paws, autumn is slow in coming. I planned the trip for this weekend to coincide with a Friday morning cold front, but it's stalled one state over, and on this day of departure for the Piney Woods of East Texas, the forecast high is 105. Jake doesn't care. He leaps into the back seat, regardless, excited. Woo-hoo! His eyes are bright; the crazy ears are in motion.

My thoughts are in motion too. Thinking like a handler is all about detail: when to stop, where to park, how to get Jake from one place to another without burning up his feet. (He has protective booties — amusing to the human, very puzzling to the dog — and the first time he wore them, they made him bust out in a kind of soft-shoe routine in the house. But his feet are so slender that ultimately Jake flips them off when he walks. They fly in every direction, some achieving remarkable distance.)

We're heading toward Louisiana, then we'll wander south along the state line before curving west-northwest back home. I double-check the tire pressure, tick off the travel checklist. We've got a map and a compass. We've got a GPS. We've got food and water for the road, a first-aid kit that works for both of us, a list of emergency vets along the way. We've got Jake's Public Access Test standards on the front seat, a spiral-bound for taking notes, and *Bless my heart, bless my soul,* we've got Alabama Shakes on the radio, telling us to *hold on.*

Eastbound then southeast-bound for mile after mile. The Dallas–Ft. Worth Metroplex falls behind us. Plains give way slowly to the first tall trees of the Piney Woods. Brown terrain suddenly shifts to a surprising green, as though some arbitrary line divides who gets rain and who does not. We turn off the major freeway and onto the first of a series of county roads and state highways that will lead us along a chain of small towns toward Louisiana.

SAR colleagues call this a Starbucks-free zone, and that's true. The biggest of the small towns have a McDonald's or a Subway. The smallest do not support even a fast-food franchise, and they are the places I like best, the ones that intrigue me most. It's been a while, but I've driven home from searches through some of these towns. Battered by time, improbably resilient, most of them are anchored by a main street lined with buildings from the 1800s, modern businesses retrofitted into leaning brick walls and rusted tin ceilings. Sometimes you find old trees and shade in the center of these towns. Jake and I stop often in order to train. He hops out of the car bright and fresh each time. A first road-trip test item checked off: Jake doesn't get carsick, even going long distances.

Unload the dog, vest up the dog, Heel, Sit, Stay, Brace. Where once-wiggly puppy Jake made a game of it every time I put a leash on him, he is attentive now. At each stop he waits for the Unload command and holds still for all the straps and buckles of his service harness, which he ignores once it's on. He waits for me to be ready without straining to get on with it. I wish I had a pant reflex. Jake's more patient that I am with the relentless heat.

Neither threatened by new places, nor overly stimulated, he seems happy to explore beside me at my pace. That pace is purposely slow. For Jake, every command on these stops is a service test item to be repeated dozens of times across the trip. I think of Puzzle, who would never enjoy this. My search dog's public etiquette is gracious, but she has always seemed resigned, acquiescing in hope that a search might follow. I look down to Jake, a different glad-to-be-here dog, whose happiness at this quieter work may be the key to his success at it.

We hug the shadows beneath awnings and peer into plate-glass windows that once framed hardware stores and soda fountains. In one town, an older man with a black Lab crosses the street not far from us. They hustle in step across the hot asphalt and head for a bench beneath a pecan tree. I think of Gene and Merlin as I watch the old dog ease down his hips and the man lean over to knead his ears like bread.

At a next stop, two children and their mama meet us. It's a nervy little moment for the mama, whose excited kids have been told not to touch the working dog, and—hands in their pockets—they're trying not to. And it's a nervy little moment for Jake, in his Sit, who loves children, and these are at his height, and yearning, with candy-smeared faces. The young mother and I hold the moment in trembling faith. Just long enough. We don't push it. "Well done!" says the young mother, hustling. "Good boy!" I say to Jake, as we walk briskly, briskly away.

Whew.

We look for black trash bags, but don't find them. We stop at one roadside stand that has kites and colored windsocks and garden whirligigs for sale. I try to imagine how a dog might frame them. It's a hot, breezy day. We walk through all the fluttering without much problem. Jake is quite interested in nylon ducks and pink flamingos with rotating wings. He ignores a fluttering ghost and a zombie in shivering rags that dangle from a tree for Halloween. These things don't faze him, but he freezes a moment in front of a spangled, inflatable pink party princess. Something about her is suspicious. He can't meet her vacuous eyes.

Back in the car, we wind down long roads from afternoon into evening, many of them roads of memory. I remember searching some of these places. There is a field we grid-walked on the space shuttle recovery. This driveway leads up the rise to that farmhouse, which is still leaning, as if peering down at the street. We pass snow-cone booths and farm stands, derelict churches, a grocery store where almost a dozen searchers went for cough drops after too much time in the sleet.

We pass rutted red trails leading into deep woods. We pass wildlife. A rabbit briefly races a grass verge beside us. Hunched turkey vultures glance up from some flat thing they have found beside the road. I glance in the rearview mirror to see how Jake continues to take the adventure. Not restless, not carsick, he is sitting up in his seat-belt harness, gazing out the window with a contented pant. He is a dog that loves the moment. Eyebrows raising, ears wheeling, his face is in motion at the curiosities blowing past us. It reads: *Interesting, interesting, wow, wow, WOW.*

Day 1:
(I write in the notebook)
Public access/task behaviors demonstrated:
*Extended car travel*
*Controlled load and unload from a vehicle*
*Controlled building entry and exit*
*Heel, Sit, Down, Stay*
*Under (at a restaurant)*
*Brace (down stairs)*
*Find Door (at a gas station)*
*Leave It (gum wad)*
*Leave It (chicken nugget)*
*Leave It (random puddle of goo)*
*Note:* startled by blow-up princess (no bark, held his ground)

We're so tired when we get to our hotel room, Jake eats only half of his supper. He leads me to the door for his nighttime constitutional, then takes us back to the right room with sleepy nonchalance. *Door (hotel room), Take Us Back (hotel room).* I'll add those notes . . . tomorrow. Jake goes belly-up by the television. To the sound of his snoring, I fall asleep in my clothes.

Day 2: Which comes first — the eighteen-wheeler or the retriever? I don't know. Plenty of car-chasing dogs have rushed us today, some on chains and some behind fences, country dogs of every description, loving their sport. And then comes this one late in the afternoon, this tubby black Lab on a porch, who spots my car approaching from one direction and a big truck merging onto our road from another. She wants to chase. She is up and heading for us, springing into her run so hard that I think the chain, when it checks her, will surely snap her neck. My God, she is fast.

But there is no chain, and there is no fence. The Lab is through the brush and heedless in the road, barking, barking, barking, and whether the driver of the merging truck doesn't see my car, doesn't care, or steers wildly to avoid hitting the dog head-on, I don't know. I lose sight of her as the truck slurs a wide arc across the pavement before us. I hear our simultaneous horns. Hitting the brakes and jerking the steering wheel, I watch the back of the truck slide so close that I can see finger streaks across the dirt of it — and at the same time I brace for collision, I expect the thud of a struck dog and its cry.

We miss. We miss the truck, which seems impossible, but the truck slings some of its load free from the covering tarp, a spray of stones. One rock strikes my windshield hard enough to crack it as we skid to the edge of a ditch. The car stops, but the crack continues its momentum, spreading upward across the windshield, curving downward before my eyes.

The truck rumbles on. My last, breathless sight of it is a sign on the back, below the tarp: KEEP DISTANCE 200 FT. NOT RESPONSIBLE FOR WINDSHIELDS.

Fumbling with my shoulder harness, I twist to look back to Jake. He hasn't made a sound. He is gamely struggling up from where his seatbelt harness has caught him, one leg twisted, his ears swung back. He looks surprised, but not distressed. I pull to a safer verge of the road and put on the flashers. If Jake's fine, we're fine.

*The dog,* I think as I get out. There is no sign of the black Lab that rushed us — not on the road and not in the ditches on either side, no blood across the pavement. She is so not there as I look for her that in the moment it almost seems like I imagined that plump, black dog skimming across the ground. It isn't until I open the door to check on Jake that I feel a dog nose against the back of my knee and hear her whimper of greeting. And here she is, a friendly dog despite her car-chasing. She and Jake stretch nose to nose, their tails swishing some common message. I look them both over. Neither is worse for the experience. The Lab would like to climb into the car. She would really like to climb into the car. She washes my face as I struggle to keep her out of it. She is adoring and brainless and, oh, she is pregnant. And lucky. She follows easily when I lead her by her collar back across the road and toward her house. Her owner meets me with apologies. She says something about bug spray in the house so the dog was outside, something about Daisy never leaves the yard.

Which is not quite true.

When I return to the car, my hands are shaking from the rush of adrenaline Jake can surely smell. We won't drive again just yet. We sit together while I massage his back and neck; he drapes his forepaws across my knees. I can feel a creeping stiffness along my right shoulder — from the tensing for impact? From the hard cut on the wheel? I'm going to be sore tomorrow. Remembering that twisted leg, I wonder if Jake will too. He's enjoying the attention. He seems comfortable enough right now.

Across the road, the capering black Lab toddles into the house with her owner.

*Calmly responds to unfamiliar dog.*

I laugh when I remember that service-test item. I tell Jake he nailed that one, and then some.

Day 3: All the way back to my first meeting with Bob and Haska, service dog handlers have told me that their worst confrontations are often in small-town stores and restaurants — a problem caused mostly, some think, by lack of exposure rather than malice. Whatever the reason, there's a crapshoot quality to every interaction, and there are handlers who still avoid local businesses altogether, trusting corporate chains to have a better understanding of the ADA. While I don't doubt their experiences, I live near a cluster of small towns and work with a service dog in training in most of them. So far it's never happened to me. Maybe we've entered a new era. Maybe I've just been lucky.

But out here, well apart from the metropolitan area, it could be different. Or not. Dogs in houses, dogs in yards; we pass a woman pushing a toddler and a poodle in a tandem stroller down a sidewalk. Resting in the shade of an awning, a sleepy pointer mix, blushed red with local dirt, hardly opens an eye when we stop for gas. We are deep in dog country it appears — but so far on this trip, I've not seen another service dog of any kind.

A café promises FAMILY FARM HOMEMADE PEACH PIE. Everyone in the restaurant looks up when Jake and I enter. I am hobbling a little from the long drive and sore outright from the yesterday's incident with the truck. Jake's in his stability harness, tagged for Service, and as I sit and gesture him Under, he folds himself beneath a table and lays his head on his paws. No one gives him much attention. The café might seat forty people on a good day. Today it seats three (none in sandals) and a dog. It smells like the L & M Café of my childhood, a comfortable blend of cigarette smoke, sugar, coffee, and grease.

A teenager in a pink T-shirt takes my order — I'm having that pie and coffee — and Jake and I are settled in comfortably when a woman elbows out from the kitchen, drying her hands. She exchanges comments with the waitress and, following a gesture of the teenager's head, looks toward us. She is a damp, flushed, pretty woman with pinned-up hair that has fallen a little loose. Her expression is a little guarded, as though she has something difficult to say and isn't sure how to say it. When she passes, one of the men drinking coffee at the counter wheels sideways on his stool, gazing out the plate-glass window with us in his periphery. Though Jake Piper is credentialed as a service dog in training, and, with provisions, local law allows training dogs in public spaces, I've heard enough about the prove-it moment that I feel my guard rising. *Here it comes*, I think, *and I haven't even touched my coffee yet.*

But when she gets to us, instead of standing at our table with her hands on her hips, the woman touches a chair and asks if she can sit. She just wanted to say hello, she says, after an awkward silence. Then she leans forward over folded arms and, looking at Jake, whispers that she doesn't want to embarrass me, she doesn't want to make a scene, but — she gestures for a bowl of water for Jake — her daughter wants to try for a scholarship that gives volunteerism credits, and she can earn some by raising a service dog puppy. Great idea, seemed to her. But they have no idea where to get one. She asks where I got mine. Do service dog groups have a Facebook page? she wonders. Could I point her someplace on the web?

I can.

Three cups of coffee and that generous slice of pie later, the woman and I are still talking, and I've stayed far longer than I intended to. Service dogs she knew, psych service dogs she didn't, but she's glad to hear about them, she says. She has seen a friend's postpartum depression so serious that she was hospitalized on suicide watch. It will take her friend a while to come back from that, if she ever really does. Folks that haven't been there have no idea how disabling depression can be.

"There are dogs for that," she says in wonder. "And some of them come from shelters. Who would ever imagine?"

We head out, the woman accompanying us. "Such a good boy," she says of Jake, who pads slowly beside me across patched and shadowed pavement. She's resisted petting him — and he, good boy, has resisted asking — but she's wrapped a hamburger patty in foil for him for later. I see his eyes squinch and his cheeks wuffle with the promise of it. We depart on that, carrying kindness, taking the scenic route she recommended rather than the road marked by the GPS.

Peace won by inches. That's how Merion measured her change at the side of a dog. Something strikes in a terrible place, and the thing you think will own you, the thing you believe you will never get past, is survived. She could have been speaking for most of the psych dog handlers I know — Bob, Kristin, Gene, Alex, Nancy, and how many others; the living respondents to disaster of every kind. In a way, she could have been speaking for me.

I think of her now as I stand in a gravel parking lot with Jake Piper. It is a familiar place, though there's a failed gas station I don't remember to one side and a restaurant with a new porch and paint job on the other. The trees are taller, the thick brush green where I remember only a winter tangle of weeds. It's a place I never planned to come back to.

For right reasons and for wrong ones, I left a suffering dog behind here once. Though I can still see her clearly, that half-starved pit bull wagging for kibble is no longer here. That was a decade ago, and she would be very old now, if she was still alive at all. There's no replacement dog either. There's no handwritten sign. I wonder about that history — wonder if the law intervened, if the house changed owners, or if those owners somehow had a change of heart. I am no better at forgiving them than I am at forgiving myself, but standing where I did in that moment of grief and divided conscience, I've put anger to rest. And guilt. I realize now that any choice I made beside that sweet, long-ago dog would have been, in some way, the wrong one. I did the best that I could. There will be dogs you can save and dogs you cannot, Paula once told me. I look at Jake and recognize how great a part that pit bull played in Jake's own saving, in the saving of every rescued dog I've since brought into the house.

Jake is happily oblivious to all this. He knows nothing of the place. He is out of the car to shake off stiff muscles and to sniff and to wander, off-duty, unvested. Ambling now to the long end of his free lead, he works his nose low, then pauses a moment and stretches. But he seems to recognize my pensiveness. Every few minutes Jake comes back to me — just a quick check-in, though I haven't asked for it. I think of Alex's description of Roscoe — *Even when he wasn't beside me, he was with me* — and that spirit is true also of Jake. In vest or out of it, he never carries his duty to a partner like a burden. Now he snorts into a patch of turf and then comes trotting back to nudge my palm and look into my eyes, like *Heyup, heyup, howzit?*

It's fine. The wind has shifted. No longer resonant, this place is just somewhere that I have been.

"Let's go home," I tell Jake. It's time. The place is getting busy. An SUV pulls in next to us, explodes grade-school cheerleaders. A dark-haired woman dashes into the restaurant and out again, her hands full of take-out boxes. We slip out of the parking lot as a truck with two hounds in the back pulls into it and then pauses nearby on the gravel.

The driver is texting. His radio's blaring Tim McGraw's "Truck Yeah" — the sad-eyed hounds' panting is perfectly in sync. They appear to be rocking out.

Let's go home. North-northwestward, we'll be homebound along new roads and back-traveling a few we came in on. It's been a good trip, I think. A fast trip; I feel it. We accomplished what we set out to do — took a long look at Jake's maturity, his diligence, and, more than that, his happiness at this work, and took a long look at myself as partner too. And we've made some new friends, dog and human. We've had some surprises. We've said some goodbyes. True of not only this trip but the whole journey, I realize, from that first moment in a Baltimore restaurant when a man said, This is my story, and this is what my dog can do.

We have full days coming. Less an ending than a rite of passage, this trip has framed what's ahead for us. Jake and I have tests to take, new service tasks to train. Mizzen and Ollie have therapy visits booked in the coming months. And Puzzle, who will at first sulk when we return home to her, has more searches to make and much more to teach me about the pathways of scent. There are good dogs out there with service hearts, too, ready to be found.

We have so much to learn.

I'm restless in the driver's seat, but Jake Piper's found the sweet spot in his safety harness, a place where he can tilt back, rest his head, and work his nose in the slipstream at will. He's worked hard. Sometimes I glance back and see him dozing and nodding like a bobble-head doll. Sometimes I see his mouth open and his eyes wide as he gazes out the window, like this is the best life ever, like he has seen the most amazing things.

# *Afterword*

IT WAS NEVER MY intention to start a service dog nonprofit. The curiosity sparked by Bob and Haska and furthered by Puzzle would result, I thought, in a lot of research, some volunteerism with one local service group or another, and plenty of advocacy for assistance dogs on Facebook and Twitter. All of those things came to pass. There are fine organizations providing beautifully trained, responsive psychiatric service dogs to clients who need them, and I never had it in mind to create another.

But for some partners, the argument for owner-training a service dog remains, and as I pursued that question — How hard is it to find a candidate dog in rescue and train it for psychiatric service? — I came to realize that parts of this experience were harder than they needed to be, that a nonprofit group to support owner-trained service dog teams could be very useful. I imagined a group that could help handlers at any point in the experience — helping would-be partners find appropriate dogs, for example, or connect with professional trainers, or make a confident air journey with their working dogs for the first time, or train additional tasks for new conditions, or acquire pet insurance. Or, at the other end of the timeline, helping handlers cope with the loss of loved canine partners or prepare future lives for their assistance dogs in the event of their own deaths.

Possibility Dogs®, Inc., came into being as a result of the many requests for help that I received while in the process of finding and training Jake Piper, Mizzen, and Ollie. We are a niche organization that focuses primarily on rescued-service-dog partnerships. We are designed to assist clients, as we are able, at any point on the journey beside their dogs. At the time of this writing, we have co-trained psychiatric service dogs beside their partners; identified dogs in rescue as good candidates for service, emotional support, or therapy; arranged for emotional support dogs and therapy candidates to interact with the public in libraries, schools, and care facilities. We have also helped first-time service dog teams make sense of the travel process by air or train in Philadelphia, Indianapolis, Dallas, Seattle, New York, Chicago, San Francisco, D.C., Baton Rouge, Los Angeles, and Burlington, Vermont. We have helped two partners locate their lost service dogs and a soldier find a home for his dog in the event that his own current medical condition cannot be survived. It's good work, and challenging.

All this said, I must agree with many of the professionals working with service dog organizations: finding a dog in rescue that is capable of becoming a thorough, reliable assistance dog is difficult. Even service organizations that have breeding programs report that as many as 40 to 60 percent of their carefully bred candidate dogs wash out or are re-careered. Certainly the number of dogs in rescue who evaluate well but are ultimately not up to the challenge of the work is equally high or higher. One dog in thirty is the statistic you often hear. I still call another professional trainer/evaluator in on every evaluation so that one of us can interact with the dog and the other can watch from a distance, with video where possible. There is value in multiple interactions and a trained second pair of eyes.

And evaluation is just the first hurdle. Some dogs do well in initial training, then reveal some health problem or emotional trigger that is profound enough to make them unsuitable for service. Sometimes such triggers can show up very late in the process. What then for the handler who has an owner-trained dog not up for the job? What then for the dog, brought home for a specific purpose? There are respondents on web forums and blog writers out there deliberating this difficult point of decision. While most service dog programs have a protocol for the dogs that do not ultimately work well, and they have a re-career or re-home path in place, for the owner-trained partnership, the handler needs to consider this possibility before the candidate dog ever comes home. (For a thought-provoking look at this, see the blog *After Gadget: Facing Life After the Loss of My Service Dog*, http://aftergadget.wordpress.com. Of particular note are posts discussing the potential washout of Barnum, Gadget's successor: http://aftergadget

.wordpress.com/2011/04/19/washing-out-make-or-heartbreak-time/.)

Have a plan B in place on behalf of the dog you have chosen. Any dog brought into the home should have a plan B, and that does not mean "If he doesn't work out, we'll take him to the pound." I mean a responsibly chosen, viable safe harbor. I would say the same for anyone simply choosing a pet — be fair, be careful, be prepared, be diligent for a creature that deserves, as we all do, a good life.

Caveats aside, the good news is that rescuing and owner-training an assistance dog can be done, and brilliantly, for both dog and handler. I would do it again without hesitation.

If I have any advice to offer about the owner-trained service partnership, it is first to absolutely collaborate with professionals on the finding, evaluating, and training of the candidate dog. Find a trainer with service experience if you can. A community of supportive peers is also a great thing. The gracious help and experiences of other handlers made my own work beside Jake so much stronger.

• • •

Readers are often curious about what happened next for the influential figures that leave their mark across a memoir. At the time of this writing, I know the next chapters for some of them.

- After more than twenty-five years working rescue, Paula retired. After much love and "thorough spoiling" in Paula's house, whistling Jasper would go on to become an emotional support dog for a chronically ill and bedridden child.
- Severe arthritis forced black Lab Merlin into retirement. Though his partner, Gene, credited Merlin for helping him heal to the point that he no longer required a service dog, Gene did bring in a rescued young poodle mix as company for Merlin, who was quickly besotted.
- Alex, challenged and changed by whippet–pit bull Roscoe, did not need another service dog. After Roscoe's death, Alex rescued a second pit bull mix, now a loved family pet and companion.
- Beautiful, soulful Lexie continues to serve beside partner Nancy, answering her needs with insistence and devotion. Nancy, her husband, and Lexie have recently rescued Pomeranian Joey, a joyful personality, a mighty presence.

- Search dog Puzzle recovered completely from the six-month skin infection that so mystified us. Working mostly with special-needs missing-person cases, at nine years old Puzzle still works search-and-rescue and assists in the training of service dog orientation tasks. She is in love with her job, happy to work her nose in any way we ask her to.

- Fo'c'sle Jack, the pirate Pom, big eater, lover of treats — the dashing therapy dog that always had an angle — died just days before this book was finished. Though not a rescue, Jack was truly my first working partner of any kind, the one who showed me what goodness a dog can bring forward when given the chance.

- Mr. Sprits'l, the small Pomeranian writ large, continues commenting on everything in the house and outside it. How did I manage, what did the world do, before Mr. Sprits'l was here to put things in order? Jobless, vestless, like so many loved pets, he's a bright spark nonetheless. I couldn't ask more of him. I wouldn't change a thing.

- In late 2008, Smokey began paying close attention to Ellen, a good friend of Erin's and mine that he has long known. Particularly in that period, if Ellen was in the house, Smokey was where she was. He followed her from room to room. He sat where she sat. Not long after we noted this new hovering, Ellen was diagnosed with breast cancer. The condition was advanced when it was discovered, the treatment rigorous and debilitating, and in the period of Ellen's long struggle through chemotherapy, aggressive surgery, and radiation, Smokey attached himself to her at every opportunity, as close as he could possibly get. We both remembered Erin. We'd seen Smokey's behavior before. Smokey gave Ellen the same attentiveness he'd devoted to his Erin, but with a difference. This time he was equally involved, but perhaps he had learned something beside his first person, or perhaps he'd just matured. This time, he was calm where he had once been overwrought. Smokey didn't interfere with caregivers or snap when he was gestured from Ellen's side. At the time of this writing, Ellen is three years cancer-free. We can't really prove why Smokey hovered near before her diagnosis, but Ellen credits his saving presence through the worst of a terrible year.

- Maddye the cat curled up in sunlight until her last day with us, in January 2012. The playful feud long over, it was Jake who led me to Maddye in the health crisis that took her.

- Jake Piper has passed his Public Access Test and trains forward on new tasks beside me — psych service, utility, mobility. In autumn 2012, I was medically advised to prepare for the need of a service dog. That Jake Piper had already begun the process sits somewhere for me between serendipity and miracle. He is both teaching partner and valued helpmate and bound to become more so.

- Chocolate Pomeranian Mizzen is now a wellness day companion through Possibility Dogs, an emotional support/therapy Pom that travels to private homes for daylong visits. Sociable and merrily adaptable, Mizzen enjoys this. She likes going out to new places, and she loves coming home again.

- Tiny, gentle Ollie T is a therapy dog that works primarily with seniors and ill children. Somewhere in the intersection of cartoon features and charm is a serenity one caregiver described as "mystical." Ollie's presence soothes.

# Acknowledgments

I COULD NOT HAVE written this book without the encouragement, information, solid support, and occasional handholding of a whole lot of people. First thanks go to all the patient folks — literally hundreds — of mental health professionals; dog evaluators and trainers; search-and-rescue colleagues; service organization personnel; service and therapy dog handlers; and animal rescue volunteers who contributed generously and mightily to every aspect of this book. Whether I have simply followed your work from afar, exchanged only a sentence with you, had a long conversation, or maintained a working relationship across years, I learned from you, am still learning as fast as you can teach me, and I cannot thank you enough. Special thanks to Dr. Joan Esnayra, Dr. Jon Morris, Suzan Morris, Fleta Kirk, Cindi MacPherson, Jerry Seevers, Susan Blatz, Darcie Boltz, Joan Froling, Susan Ruby, Emma Parsons, Melanie Jannery, Jennifer Arnold, Dr. Patricia Mc-Connell, Meg Boscov, Mary Doane, Pam McKinney, Bonnie McCririe Hale, Karen Deeds, Kim Cain, Carolyn Zagami, Sue Sternberg, Risë van Fleet, Barb Heary Gadola, Jill Blackstone, Kimberly R. Kelly, Steve Duno, Jane Miller, Steve Dale, Barbara Babikian, Jill Schilp, Susan Schultz, and Carolyn Marr.

Many thanks to my parents; my longtime friends Ellen Sanchez, Marina Hsieh, and Devon Thomas Treadwell (who said *Attagirl* when I needed it and *Oh, stop whining* when that was the better response); and the friends I have been lucky enough to make online. You know how much you mean to me and how much strength I've drawn from your support, especially in the struggle to rescue and the effort to save. Thank you for every kindness.

I am fortunate in my agent, Jim Hornfischer, and my editor at

Houghton Mifflin Harcourt, Susan Canavan, for their provocative, thoughtful insight into my work away from the keyboard and my effort to wrangle it onto the page. Thanks also to others at Houghton Mifflin Harcourt: Martha Kennedy, Tracy Roe, Carla Gray, Sarah Iani, Melissa Lotfy, and Taryn Roeder.

Though they would better understand this in toys and treats, much love and thank you to the beautiful animals that inform every day of my life. Puzzle, Jake Piper, Fo'c'sle Jack, Mr. Sprits'l, Schooner, Jemmy Ducks, Mizzen, Ollie T, Maddye, Rumblecat, Misty, Smokey, newcomer Soleil, and Sam have been beside me on this journey. Some remain, and some, much missed, have gone ahead.

A last thank you to Tricia Helfer, Jonathan Marshall, Shauna Galligan, and Mark Derwin, whose high-speed, cross-country, eleventh-hour efforts gave Little Ollie a second chance.